GROWTH AND PROBLEMS OF PRINTING INDUSTRY

HIRA KANT JHA

INTER-INDIA PUBLICATIONS
D-17 RAJA GARDEN EXTN. NEW DELHI-110015 (INDIA)

Cataloging in Publication Data—DK

Jha, Hira Kant, 1945–

Growth and problems of printing industry.

Bibliography: p.

Includes index.

1. Printing industry—India—Patna. 2. Printing—India—Patna. I. Title.

ISBN 81-210-0170-6
First Published in India in 1986

Hira Kant **Jha** (*b. 1945-*)

Published in India by:
M.C. Mittal
INTER-INDIA PUBLICATIONS
D-17, RAJA GARDEN EXTN.,
NEW DELHI-110015 (INDIA)
PHONE : 5413145

Printed by
Alka Enterprises
at Sunil Printers, Naraina, New Delhi-110028

GROWTH AND PROBLEMS OF PRINTING INDUSTRY

Dedicated

to

my reverend parents

Foreword

As invention of steam engine revolutionised production in the world, so did printing press revolutionalise communication. Shruti and Smriti (hearing and remembering) need no more be the medium of transmitting knowledge. Leaves of trees need no more be used for recording experiences and opinions. Literature need not be limited to a few manuscripts. Printing has opened up unlimited opportunities for the transmission of knowledge, information, experience, ideas and opinions on a vast scale and at tremendous speed. Revolution in its technology has accelerated this process all the more. And yet unfortunately printing continues to be one of the most unorganised activities in the world. It is, therefore, only in the fitness of things that an activity so profound in significance and yet so unorganised should receive the attention of academicians for a proper diagnosis of its problems and for suggesting solutions where possible. Since little work has been done in this field so far, Dr. Hira Kant Jha deserves special appreciation for taking up this study.

Dr. Jha had to take considerable pains in visiting a large number of printing units situated in all kinds of good and bad surroundings. In an unorganised activity like this no printed reports are available, no certified accounts are present, no organisation charts exist. There is no uniformity in principles and practices, routines and procedures. Dr. Jha had, therefore, to collect all the relevant information for this study through personal interviews and a detailed questionnaire. He has touched a number of important problems of this activity. Beginning with a brief review of evolution and growth of this activity he has passed on to a detailed discussion of its organisation, its regulation by the State, its working conditions, labour relations, capacity utilization, labour efficiency, mutual cooperation, etc. On the basis of the factual analysis of the data and information collected by him, Dr. Jha has arrived at valuable conclusions on every aspect of the problem studied by him. If this activity is to be put on sound footing, it is necessary that it should be better organised. In this connection Dr. Jha suggests the setting up of a Central Coordination Committee to identify the problems of this activity and to take effective steps for its planned development. He further suggests that proper training facilities in printing technology must be provided to improve the quality of work in this activity. These suggestions of Dr. Jha deserve immediate attention to put this activity on a sound footing.

The book breaks new ground. Being the first attempt of its kind, it has its own shortcomings and weaknesses and yet it makes a valuable contribution to the literature on the subject. I am sure it will be of use to all those who are in any way interested in printing activity and to those who have a hand in formulating policies of national development.

N.L. NADDA
University Professor and Head of the Department; Dean, Faculty of Commerce; Director, Management Programme, Patna University.

PATNA
25-8-1986

Preface

THE present study is probably the first attempt to conduct a research investigation on printing industry in an organised or systematic manner. The study includes 414 printing presses under both the private and public sectors. The work is mainly based on field study under which information was collected from primary as well as secondary sources covering the organisation, management, labour conditions, technical position and problems of development of this important industry with special reference to Printing Industry in Patna.

Printing today is a combination of manual and mechanical art. It plays an important role in a country like India which is on way to rapid industrialisation. With the new frontiers of industrial operations, printing too has to make effective contribution by increasing participation in the process of communication and flow of information. This demands a deeper study into the problems of printing and its possible expansion, particularly in a relatively under-developed State like that of Bihar which has Patna as its capital. It is with this industrial challenge that the present work has been undertaken to form the topic of a Ph.D. thesis.

Patna—The Capital of Bihar State

Patna is an old town. It has an ancient culture and civilization of Maurya. It has been adorned by two important religions of Jainism and Buddhism. It has been a place of learning since time immemorial, and it is a centre of administration. There are about 1,000 registered and unregistered printing presses in the State. Out of this figure, Patna alone has a record strength of 414 which constitutes about 42 per cent of the total number of printing presses in the State. This shows the growing importance of the printing industry in the educational, economic and social life of the people. Increasing population further improves the role of printing. At present, the representation of printing industry on the basis of population is 1 : 70,000 in the State which does not show any appreciable representation to its population of about 7 crores. The number of government and non-government offices are developing day by day and, therefore, printing industry has still further extension of market for its goods and services in Patna as well as in the State as a whole.

Though there has been a rapid increase in the number of presses in Patna, it has also created a number of problems, such as, organisation, planning, capacity utilization, labour efficiency, finance and other economic or industrial problems. This calls for an indepth study of the industrial operations with a view to identifying the right lines of its future development. Hence, the present research on 'printing industry in Patna'.

At this stage I must express my deepest sense of gratitude to my reverend teacher Dr. N.L. Nadda, M. Com., Ph.D., University Professor, Dean and Head of the Department of Applied Economics and Commerce, Patna University, for his invaluable support in conducting this research. Words cannot express my deep regards to him.

I am heavily indebted to my respected teacher supervisor, Dr. T.P. Maitin, M. Com., Ph.D., Reader, Department of Applied Economics and Commerce, Patna University, who has been the chief architect of the present research project. I also express my deep regards and gratitude to him under whose guidance I have completed this study. The present work could not have been possible without his untiring help and close attention.

I am also thankful to other teachers in the Department of Applied Economics and Commerce, Patna University, who have always inspired me in one way or the other.

Completion of a research project is a difficult task. One has to face a number of problems in course of his investigation. It is all the more difficult to conduct an empirical study of this nature. In fact, the success of the study largely depends upon the co-operation of those who actually deal with the subject. In this study I have been blessed by a large number of printers and workers, such as, Machinemen, Compositors, Operators, Casters, Binders, Artists and other workshop and office staff. I have visited almost all the 414 presses at one stage or the other. I have personally studied the problems of printers and workers. I, therefore, owe a deep sense of gratitude to all those who helped me in my task. I shall, however, consider my labour fully rewarded if the suggestions offered in this study are able to actually improve the conditions of work in printing industry.

I am also thankful to the members of Mudrak Parishad, the office bearers of Private and Public Printing Presses, and Employees Unions for their unlimited support extended to me in my research project.

I shall be failing in my duty if I do not express my thanks to Shri Narendra Bakshi, proprietor, Ramesh Printing Press and Shri Nagendra Prasad Sinha, proprietor, Jai Durga Press, who in spite of their heavy engagements always spared their valuable time to attend to my discussions.

I am also thankful to the Librarians of all the University Departments, Research Institutes, Vidhan Parishad and other institutions who have helped me with their rich materials.

Let me also express my feelings to my friends, Dr. K.C. Singh, M.A., Ph.D., Reader, Department of Labour and Social Welfare and Shri Udai Shanker Prasad, an employee in the Government of Bihar, for their continued encouragement and moral support in this academic venture.

I can never forget my respected brother Shri Bambhola Jha, M. Com., Secretary, Bihar State Co-operative Bank Ltd., Patna, who inspired me to undertake this project. My son, Sanjay Kumar has ambitiously assisted me in comparing the typed matter with manuscript.

I am also obliged to Shri Md. Waseem Ansari, who has taken pains to extend typing assistance to me and in bringing the work to its present shape. I am greatly thankful to Shri M.C. Mittal, Mg. Dir. of M/s Inter-India Publications, New Delhi, India, for his keen interest in the publication of this book.

Lastly, I am thankful to all those who have directly or indirectly blessed me in completing this painstaking assignment.

Patna

HIRA KANT JHA

Contents

List of Figures

List of Tables

List of Appendices

List of Appendices

List of Abbreviations

RBI	:	Reserve Bank of India.
BSTBPC	:	Bihar State Text Book Publishing Corporation
DRUPA	:	Democratic Republic Union of International Printing and Allied Machinery
IPEX	:	International Printing, Packaging, Paper Converting Graphic Art, Machinery and Allied Trades Exhibition
PAMEX	:	International Printing and Allied Machinery Exhibition
NSSIC	:	National Small Scale Industries Corporation

1

Introduction

PRINTING is a valuable aid to communication. It is an activity of production and a job involving distribution of thought and knowledge. In view of its growing importance in modern economy, it participates as an important industry in the process of development. It is an operation which is highly labour intensive, which combines all the latest techniques of reproduction and which affects the educational and cultural standards of the entire society.

Capital of Bihar : Patna

Patna is the capital of Bihar—the second largest State in India in terms of population, with an area of 173.9 thousand sq. km[1] and population of about 7 crores.[2] Literacy in this State is 26.20 per cent of the total population.[3] The per capita income is Rs. 995 as against the national per capita income of Rs. 1890.9.[4]

Being the capital of Bihar, Patna is the centre of administration. Out of a total population of about 30 lakhs in Patna district, nearly 39 per cent is in the literate group.[5] The population of Patna town is about 8 lakhs.[6] This population has rapidly grown over a period of time. Patna is an ancient city historically known as Patliputra. In early times the city was also known as Kusumpur or Pushpapur. It was also called Azimabad.

Being the chief centre of management for all State activities, Patna has a very wide market for services produced by printing industry. University, colleges, schools, commercial banks,

Reserve Bank, insurance companies, public corporations, Secretariat, co-operative societies, Posts and Telegraphs, Telephone, hospitals, political parties and a vast number of other organisations consume the services rendered by the printing industry. It is, therefore, obvious that a network of printing presses has spread throughout the town. They are of different sizes and their quality of production varies from one press to another. The range of their activities also affects their size and cost.

Printing : An Important Industrial Activity

"Printing is the only medium available for recording, transmitting and perpetuating human thought and knowledge to all who are able to read".[7] It is a technique which forms a basis for records, evidences and history.[8] It is one of the institutionalized forms of communication which is related to certain social institutions and groups all obtaining with the larger society.[9] It is thus not only an economic but also a social operation. It is a medium through which messages reach international markets. "Printing can hold the head of a country up in any international company".[10] Printing today, therefore, is "both culturally and commercially essential to human life and progress".[11]

Of late, printing has gained much importance in our day-to-day life. The examples are many. One can see a colour print hanging in the room, tooth paste from a printed tube, attractive cartons of packing, daily newspapers and cereals in a printed packet. We often receive advertising circulars, demand notes and other printed bills. We leave home passing through a galaxy of film posters on way to office. One deals with several kinds of printing stationery (letter heading, invoice, cards, memos, etc.); at lunch, one may choose his meal from a printed menu and pay his bill on a printed paper.

Thus, printing is an art inseparable from our daily life. It is a matter of skill. It is a technique of translating ideas into literature. Printing is a process through which ideas can be recorded, multiple copies of information can be obtained and facts may be circulated. It is a production which needs combination of men, money, machines and materials. Thus, it is an industrial activity. Printing, however, requires talent in presenting an accurate and clear impression of the matter. This needs efficient management of labour and equipment. But this efficiency depends upon a number of factors. Without adequate attention to the programme of printing operation, it may not be a profitable activity. It is thus a delicate trade which is continuously affected by market conditions as well as managerial decisions.

Nature of Printing Process

The nature and importance of printing industry may be analysed as follows. Printing—

1. works as a technological process;
2. records thoughts and action;
3. keeps pace with civilization;
4. circulates news and opinion;
5. helps sale of ideas;
6. assists presentation of facts;
7. allows distribution of information;
8. offers employment opportunities;
9. serves salesmanship and publicity;
10. acts as a medium of expression and instruction;
11. helps growth of subsidiary industries; and
12. promotes training of manpower.

Thus, printing contributes to education, communication and public relation. It plays an important role in the economic and social life of a country. It is as important as reading and writing itself. It is printing industry, therefore, which helps preservation of valuable literature in the form of books and journals.

It is in the light of such nature and importance of printing industry that we embark upon the present study. But "while dealing with the printing industry in India one has to be little pessimistic, because although the printing industry is one of the largest among the most important, it has so far failed to get the recognition it so richly deserves.[12] Yet there has been no systematic or careful study of the structure and problems of this industry". It is an anomaly which has led to a situation where no reliable statistics in respect of such basic facts as the number of printing presses and persons employed therein are available. What further complicates the issue is the fact that this gigantic industry has failed to organise itself sufficiently. Obviously this is our major limitation to analyse a situation which is so difficult to identify and then to support it with relevant facts and figures.

Objective of Study

This much about the nature and importance of printing industry. Let us now place before ourselves the objective, methodology, scope and plan of the work. Printing is a developing industry and it has to keep pace with the progress of other industries. But this industry faces divergent problems partly because of its unscientific growth and partly because of government apathy towards this industry. Therefore, a deep study into the problems of this industry is needed. The major objectives of the present study are:

1. to trace the evaluation and growth of printing industry in Patna;
2. to conduct a survey of existing printing presses in Patna with a view to knowing their size, structure, operations and conditions;
3. to identify the problems of this industry; and
4. to offer appropriate suggestions for a more productive utilisation of resources in this branch of industrial operation.

Methodology

Our task is primarily of a field study. This means a combined approach to the problems of printing industry through the use of the following methods:

1. study of relevant published materials;
2. issue of questionnaires;
3. personal interviews and discussions; and
4. seminar of experts associated with the management and control of printing industry.

Thus, the present research is an attempt to assess the working of printing industry in Patna on the basis of some first-hand survey.

Scope of Investigation

The study covers an investigation into the nature, growth, operations and problems of printing presses in Patna under both public and private sectors. A questionnaire was issued by us for the purpose of collection of data from original sources (Appendix I). But all the printing presses of Patna could not respond to the questionnaire for some reason or the other. These reasons have been explained later. The nature and scope of the present survey, therefore, has been under the following framework:

1.	Total number of presses in Patna	414
2.	Number of presses approached with the questionnaire	141
3.	Number of presses under pilot survey	28
4.	Number of presses under general survey	113
5.	Number of presses responding to the questionnaire	92
6.	Number of presses not responding to the questionnaire	49
7.	Number of presses under public sector covered by detailed investigation	3

Our first task was to obtain a complete list of printing presses working in Patna. For the collection of this information, we had two sources—one primary and the other supplementary. At first, we approached the S.D.O. (Sadar) who is the authority to receive declarations about the location of the presses and thus he was the primary source of information. There we obtained a list of 160 presses in Patna (Appendix II). But this appeared to be incomplete when we started an actual verification of the list. We actually met some of the owners of the presses whose names were not in the list maintained by the Government. So we approached the former Secretary of the Mudrak Parishad, Patna, as a supplementary source of information. He provided us an additional list of 146 presses (Appendix III). Further, a list of 108 presses (Appendix IV) was obtained from the new Secretary of the Mudrak Parishad, Patna, Bihar. We thus obtained in all a list of 414 presses in Patna from the primary and supplementary sources of information.

As it was slightly difficult to contact all the presses within the limitations of the present survey, we have applied sampling technique in our study. First of all, a pilot survey was conducted of 28 presses (Appendix V) of varying sizes through personal visit. This was a detailed enquiry in order to obtain the exact nature of their organisation and operation. Out of the remaining presses, 113 presses were taken up for a general survey (Appendix VI). Thus, a total number of 141 presses have been examined by this study. The number of presses that actually responded to our questionnaire and submitted data about their working was 92 (Appendix VII). Presses which could not respond to the questionnaire despite our repeated personal efforts may be classified under the following categories:

1.	Untraceable presses	22
2.	Presses which have closed down their business	10
3.	Sick presses	8
4.	Presses which refused to attend to our questionnaire	4
5.	Presses which have made more than one declaration	5
	Total	49

Note : A detailed list of presses falling under each of the above category appears in Appendix VIII.

With a view to acquiring familiarity with the pattern of management and nature of problems of even such presses which were not directly covered under the present survey, a personal enquiry was conducted. We visited 270 presses for this purpose. These were the remaining presses from the list of 414 presses. One major finding of this enquiry was the verification of the existence of these presses itself. The result of this investigation shows that out of 270 presses only 216 presses actually exist to continue their trade operations and the rest of the 54 presses have either been dissolved or are almost traceless. 27 presses could not be traced, 23 presses had closed down their business and 4 presses refused to give us any information (Appendix IX).

Our whole investigation into the printing industry in Patna, therefore, is based upon an indepth study of 92 presses which supplied us the required data for our investigation and interpretation. But it is only for the purpose of collection of data that the basis of our survey remains confined to 92. With a view to obtaining a complete picture of the nature and problems of printing industry in Patna, almost all presses of the town have been visited by us at some stage or the other. Not only that, we have also conducted a detailed investigation into the organisation and operations of three specialised presses under the public sector, *viz.*, Government Press, University Press and Text Book Press. The results of this investigation are incorporated in three independent chapters with a view to having a comparative study of industrial performance in the two sectors of printing presses.

It is in this perspective that we plan to examine the problem under investigation.

Plan of Work

In accordance with the objectives of the study and the research methodology adopted by us, we plan to conduct this survey of the nature and problems of printing industry in Patna under the following heads:

1. Evolution and growth
2. Presses in Patna
3. Organisation and Regulation
4. Government Press
5. University Press
6. Text Book Press
7. Working Conditions
8. Labour Relations
9. Problems of Development
10. Conclusion.

After Introduction, the first chapter will incorporate the historical developments in printing industry in India as represented by its evolution and growth in the international perspective.

Patna being the capital of Bihar has a grand historical development of printing presses since 1880. The position of such a representation of this town in the field of printing industry will be examined in subsequent chapter.

The printing presses in Patna have different organisations. Moreover, they are governed by a number of statutory provisions. Hence, we shall devote a separate chapter on Organisation and Regulations of printing presses in Patna.

This is followed by a study of three biggest presses in Patna run under the public sector—one working under the University, other under the State Government and another as a public corporation. We shall first take up the Bihar Government Printing Press, Gulzarbagh and shall examine its organisation, operations and problems. The press fulfils all the printing requirements of the State Government. Hence, a separate chapter has been devoted to make a specific study of the Government Press.

Patna University is the oldest University in the State. It has its own printing press which is the oldest press among the presses owned by the other Universities of Bihar. Therefore, working of the Patna University Press demands special attention in our present investigation. We, therefore, examine the operational problems of the University Press in another chapter and make an effort to offer suggestions for its further development.

The Text Book Press is one of the leading presses in the public sector run by an autonomous body, *i.e.*, Bihar State Text Book Publishing Corporation, Patna. This requires a separate study because of its monopoly in printing and publication of text books up to secondary level in the State. Therefore, an extensive study of the Text Book Press has been made in an independent chapter.

Adequate working conditions are essential to efficiency of operations. These conditions affect the functioning of an industrial organisation. We shall, therefore, devote another full chapter to this problem with a view to evaluating the state of working conditions in printing industry in Patna.

Labour relations is yet another important area of industrial productivity. We shall, therefore, devote our attention to this problem in an independent chapter and shall analyse the effect of these relations on printing industry in Patna. This may give us an idea as to how there has been changes in attitudes of the employers and employees.

The industrial problems are of a divergent nature. These problems affect the plans and policies of management. The purpose of the next chapter, therefore, is to analyse the major problems of printing industry in Patna as a part of our research findings. We may then be able to conclude our inquiry into the nature and problems of printing industry in Patna.

The concluding chapter shall identify the major findings of the survey and offer suggestions for a more effective operation of printing industry in Patna.

It is with this plan of work that we commence our investigation.

REFERENCES

1. Government of India., *India: 1984*, Publication Division, Ministry of Information and Broadcasting, Government of India, p. 9.
2. *Ibid.*, p. 6.
3. *Ibid.*, p. 12.
4. Government of India., *Monthly Abstract of Statistics*, Central Statistical Organisation, Department of Statistics, Ministry of Planning, New Delhi, September 1984, p. 3.
5. Government of Bihar., *At A Glance: Bihar Through Figures*, Director of Statistics and Evaluation, Bihar, 1979, p. 12.
6. *Ibid.*, p. 12.
7. Mestle, J.S. and Cordon, L. Monsen, *Photomechanics and Printing*, Oxford and I.B.H., New Delhi, 1950, p. 151.
8. Johan, L. Tarr, *Printing Today*, Oxford University Press, 1949, p. 6.
9. Parikh, R.D., *The Press and Society*, A Sociological Study, Popular Prakashan, Bombay, 1905, p. 1.
10. Johan, C. Tarr., *op. cit.*, p. 6.
11. *Ibid.*, p. 11.
12. Saifuddin, T.A., *Management Aspect of Printing Industry*, Nirmala Sadanand Publishers, Bombay, 1970, p. 1.

2

Evolution and Growth

THE object of this chapter is to trace the evolution and growth of printing industry with special reference to India.

The International Perspective

The concept of reduplication which lies at the basis of the printing technique has been known in India since ancient times. Five thousand years old pictures have been found in the excavations of Mohenjodaro. Mudra (seal) appears in Kautilyas Arthsastra (Circa 100 B.C. to 400 A.D.).[1]

The origin of printing can be traced back to China. It was the need for religious literature and pictures in connection with the propagation of Buddhism that printing was invented. Blocks were used for printing portraits of the Buddha in 650 A.D. In 1907 Sir Rurel Stein discovered a book entitled Hiraka Sutra which dates back to 868 A.D. in the carves of Thousand Buddhas in China. This is believed to be the oldest printed book in the world.

In Europe, printing with the use of movable types began in the fifteenth century although block printing was known much earlier. It is, however, difficult for the historians to state categorically if Chinese inspired the European to use printing techniques or it was the other way. Even in Europe the credit for the invention of movable types is ascribed to Gutenberg of Mainz (Germany), Coster of Haarlen (Holland), Johannes Brito of Bruges (Belgium), and Pamfilo Castalde of Feltre (Italy), on the basis of these developments, the beginnings of printing industry in different countries spread over about a century in the following order (Table 2.1).

TABLE 2.1 : **International Growth of Printing Industry**

Country	Year—Printing started
Germany	1455
Italy	1465
Switzerland	1468
France	1470
Holland	1470
Belgium	1473
Spain	1474
England	1476
Denmark	1482
Sweden	1483
Portugal	1483
Russia	1556
India	1577
China	1589
Japan	1591
Philippines	1593

Source : Ranganathan, S.R., Physical Bibliography for Librarians, Asia Publishing House, London, 1974, p. 182.

"The first book printed in U.S.A. was the Bay Psalm Book in 1640. In Latin America, however, books were printed earlier in Mexico in 1627. *Calhecismo enleugna Timonguanay Castellance* by Francisco Paneya is found in Biblioteca Mursdeana".[2]

Development in India

"The credit for preparing the first types of an Indian script goes to Joao Gonsalves, another Spaniard who accompanied Bustamente to Goa".[3] Mr. Joao Gonsalves was an expert blacksmith skilled in the manufacturing of blocks. According to Fr. Sonza, he was the first in India who made types of Malabar letters with which the first book was printed. Translations of the New Testament in various languages were printed and published by Dr. Carly in the following years (Table 2.2).

TABLE 2.2 : **Printing in India**

Year	Language printing
1801	Bengali New Testament
1802	Bengali Old Testament
1808	Sanskrit
1809-11	Hindi
1801	Marathi
1811	Oriya
1814-24	Manipuri
1815-19	Assamese
1815	Punjabi
1818	Telugu
1820	Gujarati
1820-26	Dogri
1820	Kashmiri
1821	Nepalese
1821	Marwari
1825	Sindhi
1824	Khari

Note : This shows the growth of printing in India in different languages.

Printing, however, was not known in India before 1556. It was first introduced here by the Portuguese (in Goa) for publishing biblical literature. An attempt was made to set up the first printing press by Bhimyee Prakash, a native of Western India. The printing industry emerged in India in the wake of British rule. With the expansion of the power of the East India Company the number of British people coming and settling in the metropolitan cities of India went on increasing. They had come to India with an identity of interest. Because of the same identity of interest they minimised their differences of opinion and began to live in this alien country like one community, developing an atmosphere of common culture. They started printing business to keep their common culture intact and to maintain their closer relationship.

Printing came to India in the wake of proselytization by Christian missionaries. It was some of the early British administrators who recognised printing as an instrument of unsurpassed potency for spreading knowledge among the people of India.

Printing is generally accepted to have been introduced into

the Indian sub-continent by the Jesuits, who set up a press at Goa in 1556 to produce religious works and text books in the Portuguese language. Some twenty years later another Jesuit press at Malabar produced the first works in the local language of Malayalam. "More than 200 years were to pass before the first English language press was established at Hooghlee near Calcutta and the first English language production was the Grammar of the Bengali language published in 1778."[4] But printing industry in India took its real shape in seventeenth century when a number of litho presses were established in Goa and West Bengal. Subsequently a number of presses were set up in the other main centres of commerce, like Bombay and Madras. By the middle of the nineteenth century, there were hundreds of presses throughout the country. Today the largest concentration of printing establishments is found in three major centres of Calcutta, Bombay and Madras. At other smaller cities and towns also a substantial number of presses are in operation.

Printing Under Plans

This industry has all along been given a raw deal and proof is not lacking to substantiate this argument. Our five-year plans, for instance, take no account of the printing industry. The first two plans did not give any importance to any aspects of printing industry. It was during the third plan, however, that the government recognised the importance of developing the paper industry, but that too without mentioning the role of printing industry. This is a significant omission on the part of the planning process as a sizeable portion of our paper production is meant for printing purposes. The import pattern also strengthens the general impression that the printing industry has received scant attention in comparison to the importance given to paper industry. "Out of a total import of printing machinery valued at Rs. 11.70 crores during the period 1957 to 1961, machinery worth Rs. 6.5 crores was meant for newspaper establishments and Rs. 3.5 crores for government presses, leaving only Rs. 1.70 crores for the general printing industry to buy what it needed."[5] Lack of adequate support to this important industry whose contribution to economic and educational development of the country is self acknowledged is further substantiated by the observations of the Printing Committee appointed by the Government of India in 1963. The Committee felt that the printing presses were hungry for replacements. The total requirements were estimated at Rs. 4 crores and the value of the machinery produced at home was just around Rs. 2.16 lakhs. This wide gap between the demand for equipment and its indigenous production by licensed units arrested the development of printing trade in the country.

"A survey made by the Operations Research Group in 1977 shows that there are more than 40,000 printing firms but the majority of them are very small units employing fewer than twenty people.[6] Under 10 per cent of the firms can be classified as medium size units employing between 50 and 250 people while a more handful of less than 1 per cent employ more than 250."[7]

Printing Personnel

The statistics recorded in the employment review of the National Employment Service of the Government of India for the period ending 1966 show that there are 3,431 presses in the country among the total number of 37,706 industrial establishments in the manufacturing categories of all kinds. These statistics also reveal that the number of persons employed in this industry as well as of allied nature during the same period totalled 1,39,264 against a combined industrial labour force of 42,77,295.

It can be seen from these figures that printing and publishing industry cover about nine per cent of the industrial establishments in the manufacturing category and employ about three per cent of the total industrial labour force.

It is also observed that there is a large number of smaller establishments in the printing industry than in the industry as a whole, if one were to take into consideration the number of persons employed to determine the size of the unit. The average number of employees in the printing establishment works out to 41 per unit, whereas in the industrial complex it works out to 113 per unit. It is interesting to note here that the average of the printing industry is very uncomfortably close to the minimum number required for recognition under the Factories Act. This shows that there may be a large number of units in printing industry which are working without registration.

According to an estimate made by the All India Federation of Master Printers in 1964, there were about 25,000 printing presses of varied sizes in the country, employing more than 2,00,000 persons. It further states that more than 90 per cent of these presses were such small units as to employ only five persons. The total output of printing industry in India is valued at Rs. 80 crores and the industry is estimated to be growing at the rate of 15 to 20 per cent a year.

Let us now have a broad look at the technological factors which contribute to the growth of this industry.

Printing Capacity

The industry is still predominantly in the form of letter presses, which constitute about 98 per cent of the printing firms. A recent trend in medium and large firms is towards lithography; but in the smaller firms there is little evidence of any movement towards small offset and reprographic technique. Screen process is quite widely used for print of sale work, posters and textiles; but it is generally in its most elementary form. A vast majority of letter presses are entrenched in the traditional process. In the smaller presses type is hand set usually from founders type and printing is carried out on conventional platens and cylinder process. Illustrations are usually produced from zinc-line or half tone block, although some copper etching is done for art illustrations. Medium size presses tend to have monotype installations for solid text setting and possibly linc casters if they are also having regular publications. Apart from newspapers and few other publications, nearly all the work is produced from flat forms.

Printing Installations

A unique printing firm in India is Vakils of Bombay who have the enjoyable reputation of fine art printers in every sense of the term. They are so specialized in the trade that they offer a professional design service to their customers for every occasion.

India has a few sheet fed rotary letter press installations. The one at the Saroswaty Press in Calcutta where Polygraph Victoria press is used to produce 64-page telephone directory from plastic duplicate plates is a rare example of its type.

The newspaper section is more progressive and more conscious of the need to update its technology than the other areas. There are hundreds of newspaper installations with many efficiently managed and forward looking units. There are already several well-offset installations, although not all in the main printing centres and newspaper groups are using facsimile transmission.

A more interesting development is that of the Deccan Herald group which prints by conventional curved stereotypes in Madras with facsimile transmission to Bangalore, where newspapers are printed from curved powderless etches plates on a fifty-year old Goa press and also to Coimbatore and Hyderabad where the local editions are printed by web-offset.

Litho installations are very rare. Most of the medium to large firms that are conscious of the need to develop into lithography are either tackling the job piecemeal or attempting to phase in the process to supplement or replace letter press plant. Thompson Press at Faridabad is a complete litho package installation. All the heavy printing plant and most of the light machinery have been imported often at an inflated cost due to high import duties. The government looks favourably on attempt to produce machinery at home but most advances in this field have been at the lower end of industry.

Small printing presses, such as letter press and small offset machines, are home produced, usually at prices comparing favourably with imported equivalents. Larger machines are more difficult to produce economically.

Hindustan Machine Tools (H.M.T.) produces flat bed letter press machines and sheet-fed litho presses by arrangement with Nebiolo of Italy. Unfortunately, it does not seem to be cheaper to produce these machines in India than it would be to import equivalent presses from Europe. Spare and replacement parts are very difficult to acquire as every item must be fully justified for import purposes. Most of the older printing equipment is of British origin. As is obvious, the older installations have been too strong. There must be many thousands of printing machines in India of thirty, forty and even fifty years age, but still running smoothly. The modern machinery is more like of German or East European origin. Particularly prominent is equipment from the East German Polygraph Organisation besides those of Russian and Chinese technology.

Printing Knowledge

In the field of education and training printing industry comes under Apprentice Act, 1961. The duration of apprenticeship varies from two years for trades such as hand compositors and book binders to three years for camera operators and litho printers. Each apprentice is also to take a one year full time basic training course at an approved institute, followed by in-works training. The provisions of the Act, however, have not yet been fully implemented in printing industry. If each of India's 40,000 old printing firms were to employ an average of only one apprentice, there would be 40,000 apprentices undertaking the required one year course.

Printing Institutes

At the moment, there are four major institutes of printing technology situated at Allahabad, Bombay, Calcutta and Madras. The institutes were set up in 1950 as part of government plan to promote printing education. Until that time only Madras had been offering printing courses since 1927. Although most centres provide a variety of courses relating to the needs of the local industry, the main course offered by all of them is the national three-year Diploma in Printing Technology. This is a full time course with a common first year programme for all students with an option subsequently to specialize in either letter press printing or lithography. Besides general studies regarding composition, plate making, machine printing and camera work, the curriculum also includes a study of Business Management and Administration.

In addition, some other centres have been established to cater to the need for apprentice training under the government training scheme. The duration of the course is of three years leading to the award of a Diploma. Punjab University offers a degree in journalism as well as post degree qualification in Printing Technology at diploma level. The government has been thinking to set up a high level Institute of Printing Technology to offer courses up to degree level.

Printing Technology

Like most other technologies, printing in India is a gift of the West. Yet most of the newer innovations, such as photo-composition, electronic scanners, computer type setting, plastic papers and optical character recognition are still not very much prevalent in India. The advanced printing nations have gone in for these innovations in a big way. Some of them have trickled into our country too. Computer type setting has been used by the Bombay Telephone Directory and photo composition is also catching on. An Indian newspaper uses a facsimile system to bring out its edition from a neighbouring city. Even the indigenous developments have been so fast in printing technology that sometime back an Indian firm cornered an international tender for calendar printing. And this has happened a number of times.

Yet printers in India with modern know-how could probably be counted on finger tips. Majority of the presses do not have adequate funds to venture into modernisation. Many more are not even aware of the new developments in printing technology. There are about 40,000 printing units in India employing about half million people. More than half the units, however, employ less than 10 people and barely 5 per cent employ more than 50.

Most printing machines used in India are over four decades old. Unfortunately, the indigenous manufacturing capacity for printing machines is virtually non-existent (one crore against the annual demand of around Rs. 25 crores). The technical know-how required for such machines is not available on the required scale. A printing machine requires accuracy of operation. Therefore, the problem is not only to increase the number of printing machines, it is also of printing experts or at least trained manpower. Foreign techno-how is hard to come by as foreign firms keep a watch or guard their secrets. Recently, an Indian manufacturer brought out a stop-cylinder press that copied 50 years old models. Another manufacturer who brought out web-offset machine has still to prove competitive. Technological development on a satisfactory level, therefore, has been a continuous problem for the printing industry in this country. The problem of suitable printing machines has been another limitation to a full-fledged growth of the industry.

Hindustan Machine Tools, Bangalore, produces printing machines with foreign collaboration. Foreign collaboration in the field of printing machine has stifled indigenous aspirations and thus it has often proved expensive. The choice of model of printing machines and of its quality and sophistication, therefore, are liable to suffer because of restricted import policy.

Techniques and Systems

Composing

Hand Composing: 'Composing' is a system of assembling of the types for printing purposes. The types are settled by the compositor in separate lines in a page leaded or solid. Compositor is one who composes the types. The small shallow tray in which the compositor does the work of composing is called the stick. It has a handle which can hold about 11 or 12 lines of 12-point types. The compositor holds the stick in his left hand. He picks out the types one at a time with his right hand and inserts them in the stick beginning from the left side. He does not use his eyes in picking out the types nor he shows carefulness in putting the type in the right way. Yet he puts the types in the stick at right place with the help of his fingers which are normally habituated.

Mechanical Type Setting: There are three main systems of composing—Lino, mono and photo. Let us now see their system of operation.

(*i*) *Linotype Composing*: It is one of the kinds of setting of type through mechanical process. This system of composing consists of two parts. One is composing machine which uses matrixes by keyboard operation. The second part is the casting machine. The casting machine is handled by the caster, standing at the end of the machine. In this system the equipment is used for melting lead for pumping the molten metal into the matrixes. Thus, a linotype compositor does the work of Hot-Metal composing. The linotype machine composes one line at a time. For every letter there are normally 20 matrixes or more.

Lino composing is about ten times as fast as hand-composing. A one piece line is easier to handle than a line composed of twenty or more pieces. The type face is always new. There is no redistribution problem of types since the same type is not used over and over again. No broken type will arise. The finished line of type which is called "slug" is melted and cleaned. The type metal is again ready for casting. The newspaper industry widely uses the linotype of machine for the production of newspapers. This system of composing is now being challenged by photo-composing. Linotype machine is usually used for types ranging from 4 to 14 points in size.

(*ii*) *Monotype Composing*: This system of composing has two machines. The first is the keyboard machine and the second is the casting machine. Keyboard is just like a typewriter. The person handling the keyboard is called operator. The operator by pressing the key board punches out on a spool or paper tape fed from the paper tower. Letters, punctuation marks and spaces are represented by a combination of holes. They translate the press copy into a machine readable language. When the spool is finished, it is passed on to the casting room.

In the casting process the spool is fixed in a paper tower and it rolls itself over an air pressure bar. The spool of paper is punched and finally sets the die-case in motion. This brings the appropriate matrix over the flow of molten type metal. Thus, mono type composing is hot metal composing. In this system, single letters are cast. Each letter is arranged in its line in the galley one by one. The galley is then moved up as soon as each line is completed. The entire process is automatic. Further stages of work are as in hand composing. This method of composing has all the qualities of lino composing. Besides, justification of line in mechanised proof correction does not involve resetting of entire lines. Spools can be preserved and used for further editions. Monotype composing is largely used for the production of books and periodicals.

(*iii*) *Ludlow*: This type of composing is another method of mechanical type of setting. It is used chiefly for heading types or display types. The operator sets the individual matrices by hand in a stick. The stick is placed in a casting machine where the molten metal comes into contact and produces a slug.

(*iv*) *Photo Composing*: This system of composing was developed by William C. Huepner. In the photo composing, metallic types, plates and wet ink are not used. Ink is replaced by dry powder. Pressure is not applied in the traditional way to transfer ink from type to paper. There is no contact between paper and plate. In fact the composing machine of this system is a keyboard machine. Letters are arranged in the form of images on film. The receiving surface is focused through lenses. The keyboard action unmasks the appropriate letter to project its image. All these processes of inking and composing are done automatically.

Now the process of printing starts. In the process of printing the negative or positive plates are fixed to transparent cylinders. Light shines through a slot (hole) behind the film to drive ionized "smoke" particle into the paper and create the image. Thus, film setting includes two machines. The first is keyboarding machine and the second is filming machine. The work on these two machines are done independently at different places.

Printing

"A printing press is essentially an appliance by means of which a sheet of paper is pressed against the inked printing surface, usually of the size of a form, in order to transfer the ink from the type to the paper".[8] Printing, therefore, is the mechanical multiplication of text, pictures or designs by putting ink on paper or some other suitably receptive surface".[9] In its broader sense "printing covers any method of duplicating a two dimensional design from a pattern".[10] Thus, the process of printing is combination of the following elements:

(1) The printing surface or printing image carrier (*e.g.*, type face, printing plate, gravure cylinder, etc.); and
(2) The press, the ink and the paper on other printing materials.

According to *Encyclopaedia Britannica* printing is classified under the following systems:

Letter-Press Printing: It is also called relief or topographic printing. This system uses metallic types on the press. In broader sense, it is a process by which the image is transferred from a printing surface raised slightly above the body of the type or printing plate to paper. This type or printing is mostly done by platens, cylinder or rotary machines. Generally, platens are used for short runs of relatively small sheet sizes, such as tickets, stationery and office forms. Cylinders are used for medium length runs of all sheet sizes and are especially valuable for use where better quality is demanded. Rotary is used largely by newspapers and magazines for producing big quantities of reading material quickly.

Printing Through Platen: It is the most common system of printing. It may be found in almost every small printing plant. In the process of printing under this system, the bed of the platen holds a printing surface consisting of type and relief plates in a vertical position. A flat metal impression surface (the platen which gives the press its name) holds the paper of printing. When the platen opens in a horizontal position, a sheet of paper is presented to the guider. The platen closes to a vertical position and the bed moves forward to press the inked printing surface against the paper. This makes a print. After every impression the bed permits inking rollers to pass down and up over the printing surface. The platen then opens to allow removal of the printed sheet and placement of the next sheet to be printed.

Printing Through Cylinder: It has a flat inked printing surface, similar to that used in the platen. But its impression surface which holds the sheets for printing is cylindrical. The printing surface rests on a bed that moves back and forth under the inking rollers. The cylinder rotates in exact time with the movement of the bed. During one direction of the movement of the bed, the cylinder and printing surface come into contact to produce a print. During the opposite direction of the movement the cylinder is held clear of the bed, the printed sheet is removed and an unprinted sheet is placed in position for the next print.

Printing Through Rotary: It uses both cylindrical impression surface and cylindrical printing surface. It cannot print from type or flat plates, but must have one-piece plates made with the proper curvature to fit the cylinder. The plates are made by either the stereotyping process or the electrotyping process. The printing surface and impression cylinders rotate in contact with each other; the paper passes in between to produce the print. Inking rollers are banked on each side of the printing surface cylinder to apply ink as the cylinder rotates. This method of printing is used in producing books, magazines, newspapers, catalogues, packaging and commercial printing of all kinds (*e.g.*, letter heads and sales forms).

Lithography: Lithography is also called offset, offset lithography, photo offset and planography. It is a method through which the image is transferred from a flat surface. The process of printing of different types of lithography slightly differs from each other. But generally lithographic plates are in the main made photographically. Metal type is used indirectly in the form of a point or proof that is photographed for plate making; a hand written manuscript or drawing can be photographed in the same way. In the plate making process the facsimile reproduction can be made larger or smaller than the original. Lithography is widely used in commercial books, magazines and newspaper printing. In many applications such as reprinting a letter press book after the type has become worn, it is feasible to convert to offset by photographing either the proofs from the original type or the pages from a first edition. Offset is particularly suitable for making full colour reproduction of painting and colour photographs.

Dry Offset: It is called letter set also. It combines to some extent both letter press and photo offset. In this type of printing no lithographic principles are involved. A plate similar to a lithographic plate slightly thick is etched in relief just as in photo engraving but to a shallower depth. When the plate is clamped around the cylinder of an offset press, it prints directly on the blanket which offsets the image into paper. Dry offset eliminates the need for the dampening system used in the lithographic process, and this is its major advantage.

Intaglio or Gravure: It is a process through which image is transferred from a design engraved below the surface of the printing plate. In this printing process image consists of small wells in the surface of the printing plate. Ink is carried in the wells, their depths controlling the amount of ink transferred to the paper. Photogravure includes both sheet-fed gravure printing (used mainly for short-run printing of high quality work) and rotogravure, the high speed, web-fed process used for mail-order catalogues, newspaper supplements, packaging materials and other long-run works.

Silk Screen: It is fundamentally a different kind of printing. In other words it is called screen printing. The printing image carries a stencil applied tightly against silk or some other fine fabric stretched tautly on a frame. Although the screen process is the youngest of printing methods, it utilizes the oldest principles of graphic reproduction of words and designs. Silk screen printing is preferred when a heavy opaque larger of ink (or paint) is to be applied to a poster, book cover or sheet of glass. Fine detail can be reproduced by photographic methods and half-tones of suitable pictures can also be printed.

Flexography: It was originally known as aniline printing. It is a form of rotary letter press printing in which flexible rubber plates and fluid inks are used, mainly for packaging materials. The process originated in Germany in late nineteenth century, and at first aniline dyes in alcohol were employed instead of ink. After packaging engineers took the process to the United States in the 1920s. Thin inks, consisting of finely ground pigments in a vehicle of solvent and resin were developed. Extensive use of cellophane for packaging greatly increased its importance.

Colour Printing: Colour printing can be produced by virtually all of the printing processes. Though each process has a special system of preparing the printing plate itself, the fundamentals of colour printing are the same for each. Except for the use of simple line colour, colour printing usually refers to "four colour process", though more or fewer colours may be used depending upon the quality of the reproduction or the result desired.

Other Processes: Other printing processes include: Collotype (q.v.), also called the photogelatin process, which is a planographic method of printing continuous-tone images without a screen pattern, steel die or copper-plate engraving an intaglio process that produces a raised impression or letter heads and visiting cards; and several forms of raised letter printing (imitation engraving) called thermography, virkotyping or process embossing. Using letter press methods and a slow drying ink, the printed image is immediately dusted with a resinous powder by hand or machine. When heated the powder melts and fuses with the ink to produce a slightly raised effect. These are similar to the engraved result, though usually shiny in appearance.

Xerography, sometimes classified under printing process, is actually a duplicating method often used to prepare a paper-type plate for small offset presses. It is fundamentally electrical rather than chemical or mechanical in nature. Through the principles of photo conductivity and electrostatics, xerography transfers the original image (which can be simply a typewritten sheet of a printed proof) to another sheet of paper or to a paper plate for small offset presses when a large number of copies are required. Fine powder instead of ink is electrically deposited on the paper and fused into the surface by heat.

Thus, on technical basis the types of printing can be categorised into three groups. These are:

1. *Typographic Printing*: It consists of all types of letter press printing. It is a traditional type of printing practised by most of the printing presses in Patna.
2. *Planographic Printing*: It consists of many systems of printing, such as Litho, Xerographic, Flexographic, Offset, and Silk or Screen printing. On our enquiry about the printing techniques applied by presses in Patna it was found that only 40 presses make use of this method of printing, especially presses producing Urdu literature.
3. *Intaglio or Gravure*: This type of printing is used for artistic and high yielding commercial purposes. Mostly big presses adopt this system of printing, *e.g.*, Times of India publishes its *Dharmayug* magazine by this process. Presses in Patna are unknown to this system of printing mainly because of the large investments needed to instal

such equipment and relatively restricted market for such work in this region of the country.

Printing Surface

The term printing surface is used to indicate the surface from which impression is taken. Generally, there are four types of printing surfaces. These are:

Surface for Relief Printing: In this printing the surface is made of metallic type faces standing out in relief above the surrounding parts. It covers only the type faces.

Surface for Gravure Printing: In this type of printing the surface is in the form of shallow recess in a plate. Ink of low viscosity is flowed into the recess and the surface of the plate is wiped free of ink. This system is used for printing in small quantities.

Surface for Litho Printing: The matter to be printed for litho printing is written on a certain calcareous stone or specially prepared metal surface. This system is largely used for printing of semetic scripts.

Surface for Offset Printing: In this printing, the ink is rather taken only through an intermediate surface, not by the paper directly from the letter press or the gravure plate or the litho stone. These days the surface for this type of printing is also prepared by photo composition.

This shows the technical aspects of printing industry and the different systems which are used for reproduction by modern presses. These techniques represent the subsequent stages of growth of printing industry. Let us now pass on to a study of presses in Patna in the chapter which follows.

REFERENCES

1. Marshall, Johan, *Mohenjodaro and the Indus Civilization*, (Plates C ii CXVI), London, 1931, p. 1.
2. *Ibid.*, p. 2.
3. *Ibid.*, p. 9.

4. Durrant, W.R. Penrose, *The International Review of the Graphic Arts*, Worthwood Publication Ltd., I.S.B.N., London, 1978, p. 16.
5. Saifuddin, T.A., *op. cit.*, p. 3.
6. Durrant, W.R., *op. cit.*, p. 169.
7. *Ibid.*, p. 161.
8. Ranganathan, S.R., *Physical Bibliography for Librarians*, Asia Publishing House, London, 1974, p. 253.
9. *A New Survey of Universal Knowledge Encyclopaedia Britannica*, Volume 18, I.N.C. William Benton, Publisher, London, 1968, p. 541.
10. *The Encyclopaedia, Americana, International Edition*, American Corporation, International Headquarters, 575 Letington Avenue, New York-10022, 1976, p. 590.

3

Presses in Patna

THIS chapter attempts a survey of printing presses in Patna, their historical development, technical movement, areas of specialisation, nature of operation and sources of manpower.

Establishment of Presses

Let us now examine the position of the printing industry in Patna. Patna being the capital of Bihar represents to a fair extent the historical development of this industry in the State as a whole. "The earlier Patna presses were lithography or were printed with manually type and they were printed on a hand operated press which could bring out just few small sheets per day at considerable cost; later the machine made type was used for setting bigger matrices which were operated by steam".[1] This was followed by a gradual growth in the size, number and technique of printing units in Patna.

Number of Presses

The total number of printing presses in Patna as on 31st December, 1984 was 414. This is estimated as shown in Table 3.1.

Presses which were established during 1880 to 1960 and which contribute to the history of printing industry in Patna towards its effective growth are as follows:

Initial Growth

Table 3.2 shows the evolution and growth of printing presses in Patna from their beginnings in 1880 to 1960.

Table 3.2 demonstrates the history of printing industry

TABLE 3.1 : **Number of Presses**

Year	Sources			Total
	Primary	Secondary		
	S.D.O.	Former Secretary, Mudrak Parishad	Present Secretary, Mudrak Parishad	
1880-1890	0	1	0	1
1891-1900	0	0	1	1
1901-1910	0	0	0	0
1911-1920	1	1	3	5
1921-1930	2	2	2	6
1931-1940	4	4	0	8
1941-1950	2	9	3	14
1951-1960	4	17	13	34
1961-1970	115	42	40	197
1971-1980	30	46	11	87
1981 onwards	2	24	0	26
Without dates	0	0	35	35
Total	160	146	108	414

in Patna. It shows that the Khadgavilas Press is the oldest press of the town established in the year 1880. This is known to be the first organised attempt at printing in Patna which replaced almost a century old system of Lithography.* Khadgavilas Press was established by Maharaja Kumar Ramdin Singh. At that time the major language used for books published in Bihar was Kaithi. The then Director of Public Instructions, Bihar, however, desired the publication of Books in Deonagari script. It was in this context that Kumar Ramdin Singh took the responsibility of printing text books in Deonagari script. And thus the Khadgavilas Press was the first letter press** in Patna which started printing of text books in Deonagari script. In 1900, Satyasudhakar Press was established to share the printing trade of the town. In 1912 there was another big attempt in the

*Lithography is the art of printing through scribing of Litho paper transmitting the matter on an aluminium plate by a chemical process and using that plate for subsequent impression.

**Letter press is the technique of printing through metal or wooden types and/or blocks. This may be either a hand press or a machine press.

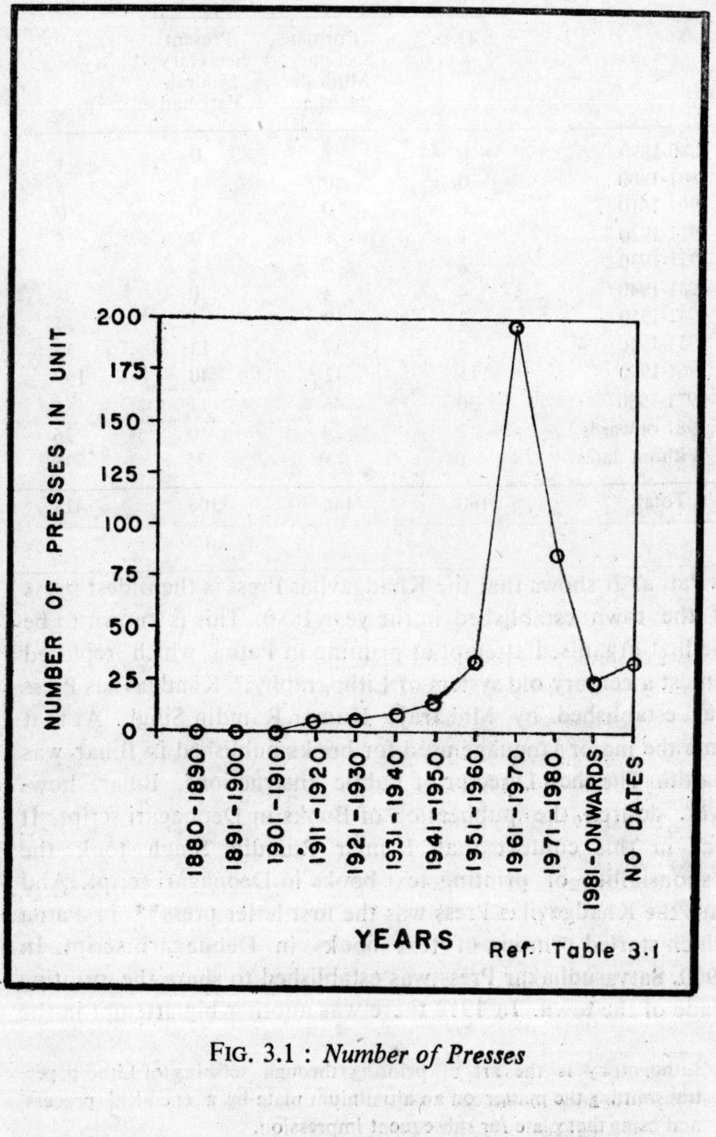

FIG. 3.1 : *Number of Presses*

TABLE 3.2 : **Printing Presses in Patna (1880-1960)**

Sl. No.	Name of Press	Year of establishment
1	2	3
1.	Khadgavilas Press	1880
2.	Satyasudhakar Press	1900
3.	The Government Press	1912
4.	Shri Shyam Bihari Press	1915
5.	Burkey Art Press	1915
6.	The Searchlight Press	1918
7.	District Board Press	1919
8.	Hind Press	1922
9.	The United Press	1924
10.	The Ramesh Printing Press	1927
11.	Jagdish Printing Press	1928
12.	Vani Press	1928
13.	The Indian Nation Press	1930
14.	Hindustani Press	1933
15.	Bharti Press	1935
16.	Bharat Printing Press and Label Industry	1935
17.	Hunkar Press	1938
18.	Patna Weekly Arts Press	1938
19.	Sri Krishna Press	1939
20.	Himachal Press	1939
21.	Union Press	1940
22.	Kalika Press	1941
23.	Litho Art Press	1942
24.	Naw Nirman Press	1943
25.	University Press	1943
26.	Spark Press	1944
27.	Prabhat Press	1945
28.	Kisan Press	1946
29.	Liberty Press	1947
30.	Ajanta Press	1947
31.	Verma Press	1948
32.	Janta Press	1948
33.	Sant Mudranalay	1948
34.	Tapan Printing Press	1949
35.	Sanjiwan Press	1951
36.	Patna University Press	1951
37.	Sarvoday Press	1951
38.	Yugantar Press	1951
39.	Vaishali Press	1951
40.	Janambhumi Press	1952

(*Contd.*)

TABLE 3.2 (*Contd.*)

1	2	3
41.	Jai Hind Press	1952
42.	Bihar Sahyogi Press	1952
43.	Bihar Co-operative Societies Press	1954
44.	Durga Press	1954
45.	Tara Press	1954
46.	Magadh Mudranalay	1955
47.	Sunil Printing Press	1956
48.	Janjagran Press	1956
49.	Gyan Press	1957
50.	Adarsh Press	1957
51.	Tarun Bharat Press	1957
52.	Shila Printing Press	1957
53.	Liberty Press	1957
54.	Hind Art Cottage	1958
55.	Vidyarthi Press	1959
56.	Manik Press	1959
57.	Jiwanjyoti Press	1959
58.	Sulekha Printing Press	1959
59.	Jaidurga Press	1960
60.	Survseva Press	1960
61.	Ghanshyam Press	1960
62.	Bhola Printing Press	1960
63.	Dharmyug Press	1960
64.	Hari Printing Press	1960
65.	Azad Hind Press	1960
66.	Shanker Press	1960
67.	Pustkalay Press	1960
68.	Deomangal Press	1960
69.	Arvind Press	1960

printing industry in Patna when the Government of Bihar established its own printing press. The District Board Press and the Burkey Art Press were started almost simultaneously in 1915. The latter, however, has been the oldest litho press of Patna. In 1918 the Searchlight Press was established which gave a daily newspaper to the State. The period from 1880 to 1920, therefore, provides important developments in the printing industry in Patna. The development continued through the expansion of letter presses for the production of text books, Government reports, religious books, journals and periodicals. The industry gained further strength after 1920.

Major Factors

An investigation into the history of printing industry in Patna reveals that the following factors contributed to the growth and expansion of this industry particularly after 1920.

1. The need for publicity and propaganda materials during the political struggle;
2. Post-Independence consciousness towards literacy;
3. Establishment of public corporations and State Boards such as the Bihar School Examination Board, Bihar Electricity Board, etc., supported by a network of several committees and public and private offices;
4. Continuous expansion of education;
5. Preparation of electoral rolls and launching of election campaigns; and
6. Production of books and journals.

Development Stages

We have seen that printing on modern lines started in Patna through the Khadagvilas Press in 1880. We shall now examine the subsequent developments in this industry by analysing its major trends of growth in Patna.

Table 3.3 records the development in the printing industry in Patna since 1914. The period represents the stages of growth of printing industry in Patna after its evolution in the last quarter of the nineteenth century.

Major Trends

A reference to Table 3.3 shows the following major trends in the development of printing industry in Patna.

1. The number of printing presses in Patna till the year 1914 was almost negligible.
2. The number of presses established between the period of the two World Wars (1915-1939), between the post-war period and Independence (1940-1947) and during the pre-plan period (1948-1951) has been almost equal.
3. The number of presses established during the First Plan period (1952-1956) was slightly low, *i.e.*, about 4 per cent.

TABLE 3.3 : **Stages of Development in Printing Industry**

Period	No. of presses established	Percentage of total
1914	1	1.0
1915-1939	8	8.7
1940-1947	8	8.7
1948-1951	8	8.7
1952-1956	4	4.3
1957-1961	11	12.0
1962-1966	23	25.0
1967-1969	11	12.0
1970 onwards	18	19.6
Total	92	100.0

4. During the period of the Second Plan (1957-1961) the number of printing presses in Patna shows an increasing trend.
5. It is during the Third Plan period (1962-1966) that the maximum number of printing presses in Patna was established contributing about 25 per cent to the total.
6. There was a decline in this rate of growth during the Fourth Plan period (1967-1969).
7. The number of presses established in Patna after 1970 is about 20 per cent of the total.

This accounts for the stages of growth of printing industry in Patna. The study shows that the number of printing presses established in Patna has fluctuated during the different periods.

Equipment Growth

Industrial development is a process of equipment growth too. It is a technical process depending on adequate plant and machinery.

Printing industry in Patna received technical assistance through various machines which were introduced in phases. Different types of equipment and years in which they were first installed have been shown in Table 3.4.

Process of Automation

As a result of equipment growth in printing industry in Patna, printers were gradually attracted towards automatic machines

TABLE 3.4 : **Equipment Growth**

	Name of Equipment	Year of introduction
1.	Hand Press	1880
2.	Treadle, Flat-Bed machines	1915
3.	Litho printing machines	1924
4.	Mono and Lino type setting (Newspapers press)	1940
5.	Rotary machines (Newspapers press)	1945
6.	Block making machines	1945
7.	Small automatic machines	1950
8.	Mono composing in the job presses	1954
9.	Offset printing machines	1960
10.	Large automatic machines	1967
11.	Photo composing (Electronic composing system)	1980

both for composing and printing purposes. The period of automation in printing industry in Patna has been during 1960-80 when the following presses installed such machines for their trade operations:

1. Government Press
2. Searchlight Press
3. Indian Nation Press
4. Himachal Press
5. Sanjiwan Press
6. Tapan Printing Press
7. Patna University Press
8. Mohan Press
9. Bihar Text Book Press
10. Janshakti Press
11. Shri Mudran
12. Balmiki Press
13. Bihar Works and Printing
14. Sunil Printing Works
15. Atma Press
16. Sidhnath Press
17. New Sahani Press
18. Jai Shri Mudranalay
19. Yugantar Press
20. Shyam Bihari Press
21. Roy Printers

Rotary and Offset

This was further supplemented by the installation of rotary and offset machines. Table 3.5 shows the years in which such machines were first installed in the presses of Patna.

Table 3.5 shows some of the leading presses of Patna who have obtained relatively modern equipment for the printing industry. It demonstrates that five presses have rotary machines and sixteen presses have offset machines.

TABLE 3.5 : **Rotary and Offset Machines in Patna**

Name of the presses	Year of establishment	Name of machines		Year of installation
		Rotary	Offset	
1. Government Press	1912	No	Yes	1960
2. Sri Shyam Bihari Press	1915	No	Yes	1976
3. Searchlight Press	1918	Yes	No	1950
4. Indian Nation Press	1930	Yes	No	1945
5. Himachal Press	1939	No	Yes	1971
6. Sanjiwan Press	1949	No	Yes	1969
7. Tapan Printing Press	1949	No	Yes	1973
8. Yugantar Press	1951	No	Yes	1980
9. Sunil Printing Press	1956	No	Yes	1982
10. Ghanshyam Press	1960	No	Yes	1980
11. Atma Press	1962	No	Yes	1982
12. Roy Printing Works	1964	No	Yes	1979
13. Sri Mudran	1965	No	Yes	1982
14. New Sahani Printing Press	1965	No	Yes	1981
15. Balmiki Press	1966	No	Yes	1982
16. Bihar Block and Works	1966	No	Yes	1979
17. Sri Sidhnath Press	1970	No	Yes	1982
18. Bihar Text Book Press	1972	No	Yes	1972
19. Janshakti Press	1973	Yes	No	1973
20. Aj Press	1982	Yes	No	1982
21. Patliputra Press	1984	Yes	No	1984

Source : Secretary, Mudrak Parishad, Bihar, Patna.

Job Specialisation

The technical growth of printing industry in Patna led to job specialisation too over a period of time; therefore, the entire printing industry consisted of specialised categories of presses. The specialisation was either of job or organisation. Some of the major categories of specialisation of printing presses in Patna are as follows:

1. Presses run by the publisher;
2. Presses run by individuals but printing for other publishers;
3. Presses for composing books;
4. Small job printers (letter pad and cash memo);
5. Big job printers (Souvenirs and Books);
6. Newspapers and magazines printers;
7. Question paper printers;
8. Register and forms printers;
9. Block and art printers;
10. Box and cartoon makers and printers;
11. Presses run by political parties;
12. Presses run by institutions;
13. Presses run by religious organisations;
14. Public sector presses; and
15. Seasonal presses.

The above classification and specialisation of presses shows that they are not confined to any one line of operation, rather different types of printing jobs have been undertaken by them under different organisational set-up.

This much about the organisational background. Let us now see the actual nature of operations of printing presses in Patna.

Time Study

Performance evaluation in printing industry is a difficult task. There is no one single criterion which can serve as an effective indicator of efficiency. The possible approaches may be to examine the quantum or the cost of output and judge its relevance under the size of the organisation.

With a view to examining the working of printing presses

in Patna, a detailed time study of three presses was conducted by us in almost all the sections. These three presses are:

1. Patna University Press,
2. Janambhumi Press,
3. Tapan Printing Press.

Patna University Press is an institutional press, while the other two presses are private presses. These presses belong to different groups and localities of Patna. Yet they have almost similar equipment for composing, printing and binding jobs.

Purpose

Our effort was to conduct a comparative study of these presses under each process of operations. The purpose was to identify the weaker aspects of the organisation, to find out the areas in which wastage of resources (including time) could be saved and to increase the overall efficiency and productivity of the different processes.

The result of the time and job study conducted in respect of (*a*) composing, (*b*) printing, (*c*) stitching, (*d*) cutting, and (*e*) ruling processes for each of the three presses is presented in Tables 3.6(*a*), 3.6(*b*) and 3.6(*c*). The figures show the average speed of work in different areas of operation and their level of activity.

Findings

The following interpretations may be drawn from the time and job studies of three representative presses in Patna:

1. There are variations in the speed of work in different sections of printing presses.
2. Tapan Printing Press, which is a private press, has the highest speed of work in all sections, *i.e.*, composing, printing and binding. This means that Tapan Printing Press has greater efficiency than the other two presses, *viz.*, Patna University Press and Janambhumi Press.
3. University Press, being an institutional unit with relatively stronger support of a State organisation, has the lowest speed of work in all sections. Slow output, therefore,

TABLE 3.6 (a) : **Patna University Press**

(*Job Study No. 1*)

Sl. No.	Job	Total coverage time		Average speed	
		Size of unit (cm)	Consumed minutes	Unit	Time minutes
1.	Composing	10 pt.—6256	808	1000 cm	129.16
2.	Distribution	10 pt.—29952	674	1000 cm	22.50
3.	Printing	—	—	—	—
	(*a*) Flat Bed	22″ × 36″ = 5600 copies	711	1000 copies	126.96
		18″ × 22″ = 6000 copies	711	1000 copies	118.05
	(*b*) Treadle	22″ × 36″ = 2000 copies	219	1000 copies	109.05
4.	Stitching	22″ × 36″ = 1000 copies	164	1000 copies	164.00
5.	Cutting	22″ × 36″ = 2000 copies	145	1000 copies	72.05
6.	Ruling	20″ × 30″ = 200 copies	235	100 copies	117.05

TABLE 3.6 (b) : **Janambhumi Press, Patna City**

(*Job Study No. 2*)

Sl. No.	Job	Total coverage		Average speed	
		Size and unit (cm)	Time consumed (minutes)	Unit	Time minutes
1.	Composing	10 pt.—6256	752	1000 cm	120.20
2.	Distribution	10 pt.—29952	560	1000 cm	18.69
3.	Printing	—	—	—	—
	(*a*) Flat Bed	22″×36″=5600 copies	690	1000 copies	123.21
		18″×22″=6000 copies	690	1000 copies	115.83
	(*b*) Treadle	22″×36″=2000 copies	195	1000 copies	97.05
4.	Stitching	22″×36″=2000 copies	119	1000 copies	59.05
5.	Cutting	22″×36″=2000 copies	110	1000 copies	55.00
6.	Ruling	20″×30″=200 copies	185	100 copies	92.05

TABLE 3.6 (c) : **Tapan Printing Press, Machhuatoli, Patna**

(*Job Study No. 3*)

Sl. No.	Job	Total coverage: Size and unit (cm)	Total coverage: Time consumed minutes	Average speed: Unit	Average speed: Time minutes
1.	Composing	10 pt.—6256	602	1000 cm	96.22
2.	Distribution	10 pt.—29952	512	1000 cm	17.03
3.	Printing	—	—	—	—
	(*a*) Flat Bed	22″ × 36″ = 5600 copies	650	1000 copies	116.07
		18″ × 22″ = 6000 copies	650	1000 copies	108.33
	(*b*) Treadle	22″ × 36″ = 2000 copies	185	1000 copies	92.05
4.	Stitching	22″ × 36″ = 1000 copies	115	1000 copies	115.00
5.	Cutting	22″ × 36″ = 2000 copies	85	1000 copies	42.05
6.	Ruling	20″ × 30″ = 200 copies	165	100 copies	82.05

demands a better efficiency and a different attitude to work.

Variations in the speed of work in printing industry, however, may be due to the following factors:

1. Nature of the work (Mathematical table, diacritical, statistical, and such other works consume more time);
2. Legibility of manuscripts;
3. Volume of job;
4. Condition of materials (quality, convenience, availability, etc.).
5. Training and experience of staff; and
6. Possible deviations (mental behaviour, environment, job satisfaction, etc.).

These factors affect both the attitude of the workers and their speed of output. However, the speed of work in different sections of the Patna University Press *vis-a-vis* other institutional presses of the town is a separate issue. This needs further probe into the speed of work in printing industry in Patna. In view of the limitations of the present investigation, it may not be possible for us to conduct time study of other institutional presses right now. When conducted on relatively wider scale the proposed study may serve as a basis of valuable results into the development of printing industry in Patna in general and of Patna University Press in particular.

Labour Market

The printing industry needs technical and skilled workers. The technological demands of this industry vary from press to press depending upon the jobs undertaken by them. There are several lines of skill-development in this industry. For example, a helper may become a distributor, then a mono caster and finally a mono operator, or in the press room an inkman may become a paper handler, then a manual platen press operator, and finally a machineman on a power operated cylinder press. The same skill development is found in the finishing and offset preparatory processes as well. Thus, printing technology requires several skill groups, such as plate preparation, making the material ready to be transferred to the plate, and the camera work associated with plate making and material correction. Each of these processes is a semi-independent activity though closer to each other. These processes are type setting, press operation or finishing.

Sources of Manpower Supply

The different kinds of labour required by the presses in Patna are drawn from three major sources. Firstly, the main channel of labour supply, is the long established tradition of non-apprenticed operatives. Each press provides some training to its workers depending upon its capacity. Most of the workers are freshers or job seekers. They offer their services as unpaid trainees. The extent to which they can acquire skill depends on the attitude of the press owners as also the mental capacity of the press worker.

The second source of manpower supply to printing industry is the Industrial Training Institute at Digha and Technical School at Gulzarbagh. These institutions run apprenticeship training courses of two-year Diploma in Printing, Composing and Binding. The trainees get a petty sum of Rs. 40 per month as stipend. Another apprenticeship training in printing technology is imparted by the Government Press, Gulzarbagh. The duration of the training is 2 to 3 years. The apprenticeship course in printing has been sponsored by the Central Government. The trainees are granted Rs. 200 per month as apprenticeship allowance (*i.e.*, stipend).

The third source is the regional School of Printing at Allahabad, Calcutta and other cities. Products of this School gain a better recognition in the market as compared to unskilled workers.

The employment opportunities available to trained personnel in printing, composing and binding are in presses both inside and outside the State. According to the Printing Apprenticeship Act, all the presses must appoint at least one trained worker for every 10 workers in the press. This, however, is seldom followed in practice by the presses in Patna.

This brings us to a close of our survey of printing presses in Patna. The study incorporates historical development, technical growth, areas of specialisation, operational tempo and sources of manpower. Let us now examine the organisation and regulation of printing units in the next chapter.

REFERENCE

1. Parikh, R.D., *op. cit.*, p. xiv.

4

Organisation and Regulation

THIS chapter analyses the organisation of printing industry in Patna and the major regulations which affect its operations.

Management and organisation have for long been considered important aspects of an industry. This, however, is an extremely complex affair. As a result of the change in management and organisation techniques and the complexities that have been identified with the progress of science, management of a printing press has become an increasingly difficult task. In general, the printing industry has been suffering from shortage of skilled labour, constant demand for higher wages and ever growing competition. This has necessitated some more effective systems of management for the efficient operation of printing presses. The systems have to be such that offer adequate profitability to the owners, on the one hand, and proper satisfaction to the workers, on the other. A printing press must be managed with some basic objective in view. In a government department or a public utility concern this objective may be to serve the society at minimum cost, or to promote regional balancing of industries, or to initiate socio-economic changes against the evils of inequalities caused by caste, colour or creed. In a private enterprise, however, adequate profitability is the motivating force behind all industrial activities. In the case of printing industry, the units have to be managed in a manner which serves as an effective channel of mass communication.

It is in this perspective that we first study the organisation and regulation of printing industry in Patna.

Organisation

A printing press may be organised under the following forms of business organisation:

(*i*) Sole proprietorship,
(*ii*) Partnership,
(*iii*) Joint-stock company—public limited and private limited,
(*iv*) Cooperative society, etc.

Table 4.1 shows the various forms of organisation and the number of printing presses working under each form in Patna.

TABLE 4.1 : **Organisation of Presses in Patna**

Form of Organisation	No. of presses	Percentage to total
(*a*) Sole proprietorship	73	79.4
(*b*) Partnership	12	13.0
(*c*) Public Limited Company	3	3.2
(*d*) Private Limited Company	2	2.2
(*e*) Cooperative Society	1	1.1
(*f*) Others	1	1.1
Total	92	100.0

Table 4.1 shows that about 80 per cent of the total printing presses in Patna are sole proprietorship concerns. Partnerships account for nearly 13 per cent and joint-stock companies just 5 per cent of the total. One press (*viz*., Sanjivan Press) is run by foreign missionaries under their own religious organisation. Thus, the printing industry in Patna shows the preference for a slightly independent form of business organisation, *i.e.*, sole proprietorship.

We shall now examine some of the characteristic features of the printing industry in Patna with particular reference to their:

Legal Provisions,
Printing Associations,
Press Premises, and
Press Machinery.

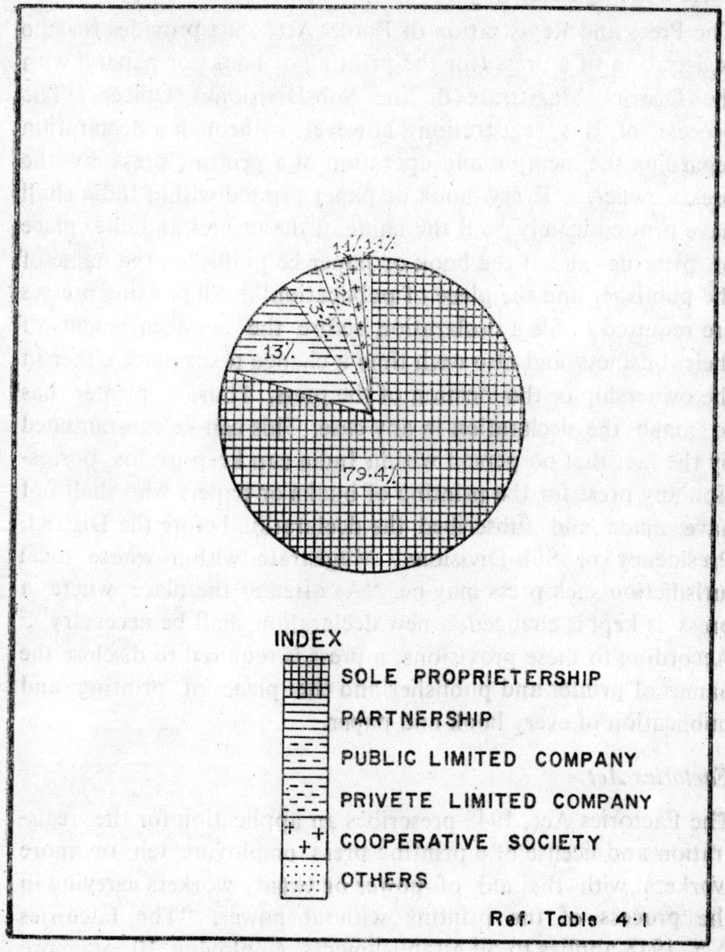

FIG. 4.1 : *Organisation of Presses in Patna*

Legal Provisions

The following three Acts affect the establishment and working of a printing presses in Bihar:

1. Press and Registration of Books Act, 1867;
2. Factories Act, 1948; and
3. Bihar Shops and Establishments Act, 1953.

Press and Registration of Books Act

The Press and Registration of Books Act, 1867 provides for the registration of a press (for the printing of books or papers) with the District Magistrate or the Sub-Divisional Officer. The process of this registration, however, is through a declaration regarding the location and operation of a printing press by the press owner. "Every book or paper printed within India shall have printed legibly on it the name of the printer and the place of printing and if the book or paper be published the name of the publisher and the place of publication".[1] All printing presses are required to file a declaration before the commencement of their business and also each time a change takes place either in the ownership or the location of the press. Thus, a printer has to make the declaration in any case. This can be substantiated by the fact that no person within India can keep in his possession any press for the printing of books or papers who shall not have made and subscribed the declaration before the District, Presidency or Sub-Divisional Magistrate within whose local jurisdiction such press may be. "As often as the place where a press is kept is changed, a new declaration shall be necessary".[2] According to these provisions, a press is required to disclose the names of printer and publisher and the place of printing and publication of every book and paper.

Factories Act

The Factories Act, 1948 prescribes an application for the registration and license of a printing press employing ten or more workers with the aid of power or twenty workers carrying in the process of the printing without power. "The Factories Act, 1948 applies to all establishments employing 10 or more workers where power is used and 20 or more workers where power is not used".[3] The application incorporates details regarding the name of the factory, its address, nature of operation, number of workers, amount of power used, etc. A press registered under this law has to follow the prescribed standards of health, safety, welfare, working hours, employment, leave, wages, occupational disease, powers of inspectors, penalties for offences, etc., as provided by the statute and adopted under the Bihar Factories Rules, 1950. This Act also lays down the conditions under which young persons may be employed on dangerous machines and prohibits the employment of women and children to certain jobs.

The Bihar Shops and Establishment Act

The Bihar Shops and Establishment Act, 1953 too makes provisions for the registration of an establishment engaged in any business, trade or profession and covers printing presses too. This needs submission of information regarding the names of the establishment, address, number of workers, management, etc. The Act regulates a press establishment regarding its opening and closing hours, daily and weekly hours of work, rest intervals for workers, duty hours of employees, weekly holidays, conditions of employment, payment of wages and overtime, etc. A printing press, therefore, has to follow the regulations of three sets of statutory provisions for the conduct of its business.

Table 4.2 shows the position of printing presses in Patna as regards their registration under the different statutes.

TABLE 4.2 : **Press Registrations in Patna**

Position	No. of presses	Percentage to total
Registered presses	64	69.6
Unregistered presses	28	30.4
Total	92	100.0

From Table 4.2 we find that nearly 70 per cent of the presses in Patna are registered under some statutes and about 30 per cent of the presses may be classified as unregistered presses. Thus, a substantial number of presses evade their registration due to one reason or the other.

The nature of this registration, however, is represented in Table 4.3.

From Table 4.3 it appears that :

(*i*) About 64 per cent of the total presses and nearly 92 per cent of the registered presses in Patna work under the Factories Act, 1948.

TABLE 4.3 : **Regulations Governing Printing Presses in Patna**

Acts	No. of presses	Percentage to total
1. Factories Act, 1948	59	92.2
2. Shops and Establishment Act, 1953	4	6.2
3. Factories and Shops and Establishment Act	1	1.6
Total	64	100.0

(*ii*) Nearly 4 per cent of the total presses and about 6 per cent of the registered presses in Patna have obtained registration under the Bihar Shops and Establishments Act, 1953.

(*iii*) Roughly 1 per cent of the total presses and about 2 per cent of the registered presses in Patna are governed both under the Factories Act, 1948 and Bihar Shops and Establishments Act, 1953.

This means that about 36 per cent of the presses in Patna are not covered by the provisions of the Factories Act, 1948 and nearly 96 per cent of the presses operate outside the purview of the Bihar Shops and Establishments Act, 1953. We thus find that the maximum number of registered presses are governed under the Factories Act, and the Bihar Shops and Establishments Act regulates a relatively small percentage of presses in Patna.

Printing Associations

The printing industry in Patna has an Employers' Association known as Mudrak Parishad. Our investigation shows that out of 414 presses in Patna only 144 presses are members of the Mudrak Parishad. This means that about 35 per cent of the presses of the town are members of this association. Out of the 92 presses covered under the present study, however, 78 presses or about 85 per cent of the total presses are the members of the Mudrak Parishad. The rest probably either do not believe in such associations or do not subscribe to the views of the Mudrak Parishad.

Press Premises

As regards the nature of premises occupied by the printing industry in Patna, our enquiry reveals that only 33 per cent of the presses have their own premises. The rest of them function in rented houses. In terms of area occupied by the printing presses, however, the position is as follows (Table 4.4).

TABLE 4.4 : **Area-wise Distribution of Presses**

Area in sq. ft.	No. of presses	Percentage to total
Below–1000	32	34.8
1001–5000	52	56.5
5001–10000	6	6.5
10001–15000	—	—
15001–20000	—	—
20001–25000	—	—
25001–30000	—	—
30001–35000	—	—
35001–40000	—	—
40001–45000	—	—
45001–50000	—	—
50001 and above	2	2.2
Total	92	100.0

Table 4.4 shows that the majority of the printing presses in Patna have a working area between 1001 and 5000 sq. ft. About 35 per cent of the presses, however, have an area below 1000 sq. ft. There are only two presses which have a relatively bigger space exceeding 50,000 sq. ft. They are Sanjiwan Press at Digha and the United Press at Bari Path (near Hathwa Market). Thus, it appears that most of the printing presses covered under the present study have smaller areas of operation. The average area occupied by the printing presses in Patna works out to about 3,940 sq. ft. per press.

Press Machinery

The size and the working capacity of the printing presses is governed by the number and variety of machines available to the trade. This position has been represented through Table 4.5.

TABLE 4.5 : **Printing Machines in Patna**

Category	No. of presses	Percentage to total
Where there is no machine	4	4.3
One machine	30	32.6
Two machines	22	23.9
Three machines	18	19.6
Four machines	10	10.9
Five machines	4	4.3
Six machines	2	2.2
Seven machines	1	1.1
Eight machines	0	0.0
Nine machines	0	0.0
Ten machines	0	0.0
Eleven machines	1	1.1
Twelve machines	0	0.0
Total	92	100.0

Table 4.5 shows the position of printing presses in Patna in terms of number of printing machines.

From a reading of Table 4.5 we find that: (*a*) the maximum number of printing presses in Patna have just one machine. This accounts for about 33 per cent of the total presses; (*b*) some of the presses work even without machines; (*c*) as the number of machines increases, the number of presses falling in that category declines; and (*d*) there is only one press which has as many as eleven machines at its disposal. This is Tapan Printing Press at Machhuatoli.

As regards the types of machines used by these presses, the position is given in Table 4.6. (also see Fig. 4.6).

It appears from Table 4.6 that a larger number of printing presses in Patna (*i.e.*, about 58%) have their own Treadle machines.* Of the 92 presses, there are 80 presses in this category. 44 presses or about 32 per cent of the total presses, however, have Flat bed machines.** 10 presses out of 92 presses

* Treadle machine: It is a type of small printing machine which is used for printing of short runs of relatively small sheet sizes.

**Flat bed machine: It is also a type of printing machine which has flat inked surface and holds the sheet for printing cylindrically.

TABLE 4.6 : **Types of Printing Machines**

Types	No. of presses	Percentage to total
Flat	44	32.0
Treadle	80	58.0
Litho	1	0.7
Mono	1	0.7
Rotary	1	0.7
Offset	10	7.2
Others	1	0.7
Total	138	100.0

have installed Offset machines which constitute 7 per cent of the total presses. The number of presses having Rotary, Mono and Litho machines is just one under each of these categories. Some of the presses have printing machines of more than one category too, *e.g.*, United Press, Gyanpith Press, Tapan Press, Sidhnath Press, Atma Press, Ghanshyam Press, etc.

Our enquiry regarding the position of Ruling machines used for assisting the printing trade in Patna shows that nearly 91 per cent of the total presses do not have any ruling machine. There are just 9 per cent of the presses in Patna which have one ruling machine for each of them.

Similarly, our survey regarding Stitching machines operating in the printing presses of Patna shows that about 58 per cent of the presses do not have any such machine. Only about 39 per cent of the presses possess one stitching machine for each of them and nearly 3 per cent of the presses have two stitching machines for each of them.

Besides Ruling and Stitching machines, there are a few other types of machines also which are used by the printing presses. These include machines for perforating, cutting, etc. Table 4.7 shows the position of the presses in terms of such subsidiary machines.

From Table 4.7 we find that about 30 per cent of the presses do not have any subsidiary machines like cutting and perforating. Each of about 56 per cent of the presses have between one to three such machines and about 14 per cent of the presses have between 4 to 10 machines for their trade operations.

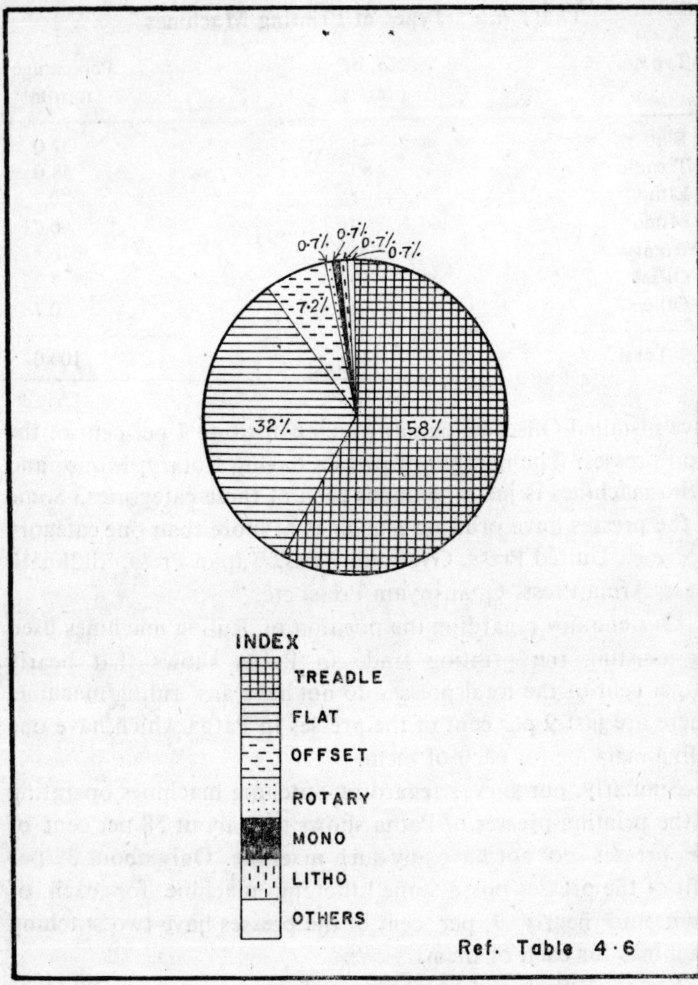

FIG. 4.2 : *Types of Printing Machines*

This presents some of the major aspects of printing industry in Patna in respect of their regulations and organisation. Let us now conclude this analysis.

Conclusion

This statistical survey of printing presses in Patna gives us the following conclusions:

TABLE 4.7 : **Subsidiary Machines for Printing Industry in Patna**

Number of machines	No. of presses	Percentage to total
Where there is no machine	28	30.4
One machine	14	15.2
Two machines	16	17.4
Three machines	21	22.8
Four machines	6	6.6
Five machines	4	4.3
Six machines	1	1.1
Seven machines	1	1.1
Eight machines	0	0.0
Nine machines	0	0.0
Ten machines	1	1.1
Total	92	100.0

Note : Subsidiary machines include perforating and such other machines used by the printing trade.

1. The corporate form of organisation has been adopted only by a negligible percentage of presses in Patna. A majority of them work under the sole proprietorship form of organisation.
2. The maximum number of presses were established in Patna during the period 1962 to 1966.
3. The majority of the presses in Patna are governed by the Factories Act, 1948. It is this Act, therefore, which provides statutory control to the printing industry in Patna.
4. Not all the printing presses of Patna are members of the Mudrak Parishad, though this is the only recognised body for the printers in Patna.
5. About 67 per cent of the presses in Patna function in rented premises.
6. This limits the area of operation up to 5000 sq. ft. for about 57 per cent of the presses.
7. The working capacity of the presses in Patna is largely represented by two types of printing machines—Treadles and Flat Bed—which they have for their operation.
8. Only a few presses have technically developed equipment like Rotary, Mono and such other machines for the

conduct of their trade. However, about 8 per cent of the total presses in Patna have installed Offset machines for their operations.

9. Only about 9 per cent of the presses have their own Ruling machines.
10. Stitching machines, however, are possessed by nearly 39 per cent of the presses.
11. Other subsidiary machines for the printing job also are not available to about 30 per cent of the presses in Patna.

These are some of the major trends of printing industry in Patna. Coming out of our investigation into this field of operation these represent the limitations under which the trade has to function. These also make one think about the problems of this trade. But more about such problems later on. For the present, we examine the organisation and operations of three different printing presses under the public sector in the chapters which follow. We shall first take up the Bihar Government Printing Press, Gulzarbagh, Patna.

REFERENCES

1. Sen, P.K., *The Press, Publications and Copyright Laws of India*, M.C. Sarkar & Sons Private Ltd., Calcutta, 1958, p. 7.
2. *Ibid.*, p. 8.
3. Saifuddin, T.A., *op. cit.*, p. 235.

5

Government Press

THE object of this chapter is to examine the organisational set-up of the Bihar Government Printing Press, Gulzarbagh, Patna, which is a public sector unit.

"With the creation of the Province of Bihar and Orissa in 1912, the Government Press at Shillong (Assam) was removed to Ranchi where the capital of the new State was temporarily located; later it was shifted to Gaya and ultimately to the premises of the old opium factory at Gulzarbagh".[1] The initial total capital investment in the press was about a crore of rupees.

Historical Background

Before 1912 Bihar and Orissa were integrated with the Bengal State. There was a well established press in Dacca, now a city of Bangladesh, where the printing works of these integrated States were done. Later on an independent printing press was opened in Calcutta by the West Bengal Government. The printing and publication works of the Government of Bihar was also done in Calcutta. Before its partition from Bengal, the printing works of the integrated States of Bihar and Orissa were undertaken by the Government Press situated at Shillong. It was in the year 1912, after the separation of Bihar and Orissa from Bengal, that a printing press was established by the Government of Bihar at Ranchi. This was known as "Printing and Stationery Stores and Publications". After few months another printing press of the Government of Bihar was brought into existence at Gaya. These two presses, however, were not properly equipped. Printing and publication requirements of the Government were increasing day by day. Therefore, the expansion of the Government Press with all modern equipment had become essential. This resulted in the establishment of a new printing press at Gulzarbagh (Patna) in the same year, *i.e.*, 1912. The press was formed as Department of Printing, Stationery Stores and Publications. At that time there were about 310 employees in the press. The Superintendent of the Press was assisted by one Deputy Superintendent, two Assistant Superintendents and one Administrative Officer. In 1936, Orissa was separated from Bihar and the State of Orissa started its own printing press.

In the year 1955, the printing section of the press was separated from that of stationery stores and publications and both the sections were put under a separate Superintendent under the overall control of the Finance Department in the Secretariat. The number of workers now increased to 675 out of which 500 workers were technical hands and the rest 175 were ministerial. The present number of press employees is 673. The press has one Superintendent assisted by the Deputy Superintendent and four Assistant Superintendents. Besides, the establishment consists of an Administrative Officer and a Labour Officer to manage the respective areas of responsibility.

The press is duly registered under the Factories Act. It is working in an area of 2,72,000 sq. ft. on the bank of the river Ganges. The main purpose of the press is to cater the services of printing and publication to the Government of Bihar and its allied departments. It has two main branches located at Ranchi and Gaya and two small branches working in the Old Secretariat and Raj Bhawan premises. These branches meet the printing requirements of Government offices on local basis and share the workload of the main press. The functions of the press are governed by the rules of the State Government described in the Government of Bihar Secretariat Instructions, 1952.[2]

Organisation Structure

Nature of Organisation

The Bihar Government Printing Press (Gulzarbagh) is a truly Government Press. It is also known as Sachivalaya Mudranalaya.

The press is looked after by a whole time Superintendent, who works under the guidance and supervision of the Joint Secretary, Department of Finance, Government of Bihar. Moreover, the Director of Printing and Stationery holds the responsibility of printing and publication jobs of the Government. The press, therefore, is controlled by the Director of Printing, Government of Bihar. The press is supported by the two sets of staff, first is office staff and second, workshop staff. We shall study the two categories of staff separately in a subsequent discussion in this chapter.

The following is the organisation structure of the Government Press.

Government of Bihar

1. Chief Minister
2. Finance Minister
3. Finance Commissioner
4. Joint Secretary, Finance
5. Under Secretary, Finance
6. Director of Printing and Stationery

Press

1. *Office*	1. Superintendent 2. Assistant Superintendent 3. Labour Officer 4. Administrative Officer 5. Section Officer 6. Assistants
2. *Workshop*	1. Foreman, Selection grade (overall in-charge) 2. Foremen 3. Sectionholder
(a) *Printing Section*	1. Foreman 2. Sectionholder 3. Jamadar 4. Machinemen 5. Shramik, Coolies, Fly Boys, etc.

(b) *Composing Section*

(*i*) Hand composing		1. Foreman 2. Sectionholder 3. Compositors
(*ii*) Mono composing		1. Foreman 2. Sectionholder 3. Operator 4. Caster 5. Stand keeper
(*iii*) Lino composing		1. Foreman 2. Sectionholder 3. Compositor 4. Stand keeper 5. Operator
(c) *Melting Section*		1. Sectionholder 2. Meltor
(d) *Correction Section*		1. Section Head 2. Corrector
(e) *Distribution Section*		1. Head Distributor 2. Distributor 3. Shramik
(f) *Mechanical Section*		1. Head Mechanic 2. Mechanic 3. Helper
(g) *Offset Section*		1. Foreman 2. Mechanic Jamadar 3. Machinemen 4. Binder
(h) *Photography Section*		1. Cameraman 2. Assistant Cameraman
(i) *Reading Section*		1. Head Proof Reader 2. Reader 3. Copyholder

(j)	*Store*	1. Store-keeper 2. Store Accountant 3. Assistants
(k)	*Paper Issue Section*	1. Godown Incharge 2. Assistants
(l)	*Gate*	1. Head Time Keeper 2. Time Keeper 3. Darwan

Procedure of Recording Work

The press maintains various records for the job done. Some of the important records maintained by the press are:

1. General register (year-wise)
2. Job register
3. Docket register
4. Receipt and Despatch register
5. Challan register
6. Bill register
7. Ledger book
8. Work maintenance register
9. Job process register
10. Job progress register

So far as the procedure of recording work is concerned the position of the Government Press is better than that of private presses. But when compared with other public presses like Patna University Press or Bihar Text Book Press, it does not show any appreciable development.

Financial Management

Capital Structure

The Government Press is wholly dependent on the Government of Bihar for its financial needs. Therefore, the financial transactions of the press are controlled by the Government of Bihar itself. The press has a total capital investment of Rs. 100 crores. It has neither equity nor borrowed capital. Financial discipline is maintained through annual budgets prepared by the press authorities and approved by the Government. The total allocation for the press during the year 1984-85 was to the tune of Rs. 2,83,10,900. Being a public enterprise the press runs its business on 'no profit no loss' principle.

Basis of Payment

Payment to the staff is made by the press on the following basis :

1. Printing	Salary/Daily wage
2. Composing	Salary/Daily wage
3. Binding	Salary/Daily wage
4. Proof Reading	Salary/Daily wage

There are 673 staff in the press. All of them are whole-time permanent staff. They are paid on monthly basis. The terms and conditions of their service are governed by the rules of the Government of Bihar. The press pays its casual workers at the rate of Rs. 6 to Rs. 10 on daily basis. Work is not done in the press on contract basis.

Price Policy

Price policy of an industry is determined by the system of costing its products. Therefore, a sound principle of costing is a prerequisite to an effective pricing policy particularly in a printing industry.

The Government Press, however, does not follow any set principles of either costing or pricing. The sole purpose of the press is to provide printing services to the Government and its different departments. Therefore, the press is not very serious about a profitable price structure. However, the press includes the following elements in its method of costing for goods and services as a part of its price policy.

Elements of Printing cost

1. Paper........................
2. Materials........................
3. Labour
   (*a*) Establishment cost...............
   (*b*) Process cost........................

(*c*) Printing..............................	
(*d*) Binding..............................	
Total..............................	
Total..............................	
Cost of storage	....
Cost of obsolescence	....
Total cost	....
Profit : 25 per cent on cost (on outside jobs)	
Total charges	....

Source : Material and costing section, Government Printing Press Gulzarbagh, Patna.

Although the Government Press is not a profit earning institution, it charges a profit of 25 per cent on outside jobs which normally come from R.B.I., Electricity Board and other government or semi-government establishments.

Technical Position

Machinery

The number and types of machines used by the press are shown in Table 5.1.

System of Composing

The press uses almost all types of composing, *viz.*, hand composing, mono composing, lino composing, ludlow composing and photo composing. Thus, the press has developed both manual and mechanical systems of composing. Photo composing through electronic process, which is a latest development, is a special feature of the Government Press. A lion's share of the composing job is done through mono composing in which the highest number of compositors are engaged. Inter-type is the only system of composing which has not been adopted by the Government Press.

Source of Supply of Machines and Materials

The Press Superintendent has been given very limited powers to make direct purchases. In exceptional situations, however, he may buy some subsidiary machines, like that of envelopes

TABLE 5.1 : **Machines**

Name of Machines		Number	Size (cm)	Speed per hour
1. *Printing Section*		31	—	—
Treadle				
(*a*) Hand feeding	4		12×16	1,000
(*b*) Automatic	8		13×19	3,000
			15×18	3,000
	12			
Flat Bed				
(*a*) Hand feeding	2		61×86	2,000
(*b*) Automatic	10		61×86	8,000
	12			
Offset				
(*a*) Romayr	4		61×86	4,000
(*b*) Platen	2		31×86	6,000
(*c*) Web, Double colour	1		61×86	15,000
	7			
2. *Mono Section*		20		
(*a*) Mono key-board	10		—	—
(*b*) Mono casting	10		—	—
	20			
3. *Photo Composition* (*Film machine*)		1	—	2,500
4. *Ruling Section*				
(*a*) Hand feeding		1	31×86	500
5. *Others*		26		
(*a*) Cutting	11		61×86	10,000
(*b*) Stitching	6		—	3,000
(*c*) Folding	5		61×86	4,000
(*d*) Drilling	2		—	—
(*e*) Pasting	2		—	—
Total		79		

Note : The above table shows the different types of machines available in the Government Press, and their number and capacity.

cutting, welding, litho, etc. For other costly machines, for printing and composing purposes the press has to depend upon the Government. It is only after the final approval of the Government that the press can purchase such machines through the recognised agencies at Bombay, Kolhapur, Delhi and Calcutta. But most of the machines used by the press are imported from foreign countries, *viz.*, U.S.S.R., U.K. and Germany.

Metals and chemicals up to Rs. 500 may be directly purchased by the press authorities from the local markets. But for larger purchase requirements the press needs Government approval and sanction. The Government sanctions a lump sum to the press for this purpose. Chemicals are purchased from Calcutta and types are manufactured in the press itself through its supercasting machines. The important types required by the press for further prints are retained and the remaining ones are melted for subsequent new types.

The press occasionally buys papers from the local market. But normally the Government grants the required quantum of paper to the press through the Director General of Supplies and Disposals, Government of India, New Delhi. This is done under the quota system of the Government direct through the paper mills. The purchase account is sent to the Accountant General, Bihar and to the press authority also—the latter being responsible for final payments.

Variety of Operations

The nature and type of jobs done by the press is determined by the Government. As the press is often over burdened with the Government work, outside jobs are rarely accepted. Yet the job register of the press shows that on an average jobs worth Rs. 5 lakhs have been given to the private presses of Patna every month during 1983-84 and 1984-85. The press is also responsible for the publication of the Government gazetteer, Government budget, Bihar Secretariat Instructions, *Bihar Samachar,* Bihar Information, Shramik, etc. The number of these publications and periodicals varies from 2,000 to 10,000 copies of every issue. Besides, the press has also to manage the printing jobs relating to ballot papers, bills, letter pads, etc., of the Government. For publishing gazetteer and budget, the press does not charge any thing from the Government. It is the Finance Department which bears the cost of these two jobs. For the rest of the work, however, the press charges according to the rates fixed by the press authorities in consultation with the Joint Secretary, Finance Department, and Director of Printing, Government of Bihar.

Labour Conditions

The staff structure of the Government Press is shown in the following tables.

TABLE 5.2 : **Number of Workshop Staff**

(1) Compositors		
(*a*) Hand	60	
(*b*) Machines	72	
(*c*) Corrector	64	
(*d*) Distributor	7	
		203
(2) Machinemen		53
(3) Fly Boys		29
(4) Binders		96
(5) Coolies		95
(6) Others		80
Total		556

TABLE 5.3 : **Number of Office Staff**

(1) Managing Director	—
(2) Superintendent	1
(3) Deputy Superintendent	1
(4) Assistant Superintendent	4
(5) Manager	—
(6) Accountant	75
(7) Press Assistant	—
(8) Typist	13
(9) Darwan	11
(10) Others	12
Total	117

From the foregoing staff structure it is clear that the Government Press has a total of 673 staff, 556 for technical jobs and 117 for office administration. This is sufficient enough to help a smooth functioning of the press work. The composing section of the press is represented by a good number of machine compositors which includes mono operators, mono casters, lino operators and casters. The total number of this category of compositors comes to 72 which is more than the number of hand compositors. It demonstrates a good trend towards mechanisation in the process of composing. Its binding section includes binders and packers of senior and junior grades. The press has different categories of workers, like roller casters, press men, galley proofmen, foremen, meltors, impositors, type suppliers, phras and firemen. The press has a substantial number of other staff too in both workshop and office to supplement the regular staff. Photographers, plate typist, labour officer, administrative officer, security officer and readers, revisors are some of the personnel working on permanent basis.

Working Conditions

Health and Safety: The press has a good location in a clean environment. Moreover, the press has made arrangements for health, safety, cleanliness and welfare facilities. Its position in other areas of working conditions are as follows :

Rest House : The press has been working double shift (day and night) during the last several years. But there is no arrangement for rest room, within the press premises for its workers. The day shift starts from 10 A.M. and ends at 5.30 P.M. The night shift starts from 5.30 P.M. and ends at 12.30 A.M. If a worker has an accident or any other untoward happening, the press canteen is used as a rest house. Though the Bihar Government Press Employees Association too has been requesting the Government for a rest house, no step has been taken in this regard uptill now.

Dispensary : Lack of medical aid to the workers of the Government Press is another problem. Though there is a public hospital near the press, the workers do not receive any special facility there. On personal enquiry from the workers of the press, we came to know that the workers of composing, melting, casting and printing sections have to be regularly examined by the Medical Officer of the hospital to know whether or not they suffer from any occupational diseases like lead poisoning, TB, etc. Before 1970 the workers were normally checked by a set of medical officers every six months. On finding any worker suffering from an occupational disease, the workers were declared unfit and they were granted six months special leave for their treatment. During the period of their treatment they received special medical aids in addition to full salary.

After 1970, however, doctors' visits to the press have been rather too poor. The workers are not being seriously examined and the press authorities do not seem to be conscious of their responsibilities in this connection. The Government Printing Press Employees Association, a union of the press workers, has urged upon the press authorities on several occasions to make suitable arrangements for medical aid to the workers within the press premises on some regular basis.

Duty Hours : The press starts working from 10 A.M., but almost all the officers of the press come to office around 10.30 A.M. There has been large number of casualties to the workers between 10 A.M. and 10.30 A.M. But on account of non-availability of officers during this period the suffering workers had to remain in the office in senseless condition. Similar is the case with night shifts. The Superintendent gets Rs. 200 as night shift allowance. But the entire night shift is left at the mercy of the night shift foreman. The Government, therefore, should take a serious note of the general apathy of the officers of the press and should take adequate steps to meet the genuine needs of the workers.

Service Conditions

Nature of Employment : The majority of staff employed in the press is of permanent category. The press has neither temporary workers nor workers on contract. In case of urgency of work, the press engages some workers on daily basis. But the number of daily wage workers has not been more than 240 in a year. These workers are neither piece-rate nor contract workers. They get their fixed payment according to a prescribed volume of work done by them. Their wages vary between Rs. 6 and Rs. 10 a day. This category of workers does not enjoy any other benefit like leave or medical facilities.

Leave Rules : The leave rules in the press are the same as applicable to the employees of the State Government. In other words, the press staff also get earned, casual, medical and extraordinary leave. This is in addition to the usual gazetted holidays granted to the Government office staff.

House Allowance : House allowance is granted to the press staff as per the rules of the Government of Bihar, *i.e.*, 15 per cent of the basic salary as paid to other employees of the Bihar Government. Out of 673 press employees only 15 workers have been granted quarters facilities. This constitutes about 2 per cent of the total number of workers in the press. This shows that the Government is probably not aware of the housing problems of the press workers. This directly affects their efficiency to work. On enquiry about the actual number of quarters available to press staff, we gathered that out of 19 residential quarters allotted to the press staff, 3 have been under the occupation of gazetted officers of the press and one remained under the control of a teacher of the Government High School for many years. All the press workers are required to report for their duties at 10 A.M. (sharp). If they are late on 4 days in a month their one day leave is deducted from their total entitlement. It is imperative, therefore, that the Government should make arrangements for suitable accommodation to the workers near the place of the job. This demands an affective housing policy of the Government for the press.

Penalties and Offences : In case of penalties for offences warning is the usual system. Bad record, wage deduction, leave cut and show cause notices are also practised by the press authority when the workers commit offences. Proceedings followed by dismissals or suspension of the workers are in practice, provided the offence is proved by the Enquiry Committee constituted by the press authority. Thus, the press authorities adopt several disciplinary measures against the offences of the workers.

Overtime Work : The Government Press works overtime occasionally, especially at the time of State and Lok Sabha elections. Overtime work was not needed before 1956. The frequency of overtime work in the press is about 6 hours per day. Mostly the permanent staff are engaged for this purpose. When there is urgency of the work, workers are also engaged on daily basis. According to the Factories Act, the overtime work has to be paid for at double the ordinary rates. But this method of payment is not practised by the press. Rather the Government Printing Press has its own systems of payment for overtime work. Payment for overtime to the permanent staff is linked with the basic salary plus dearness allowance and city allowance. The methods adopted for calculation of payment for overtime are as follows :

Permanent Staff

$$\frac{\text{Basic pay} + \text{Dearness allowance} + \text{City allowance}}{\text{Total No. of hours in working days of a month}}$$

$$\text{Suppose} = \frac{\text{Rs. } 500 + \text{Rs. } 200 + \text{Rs.} 10}{25 \text{ days} \times 6\frac{1}{2} \text{ hours (daily)}} = \frac{\text{Rs. } 710}{162\frac{1}{2} \text{ hours}}$$

$$= \text{Rs. } 4.34 \text{ per hour per day}$$

OR

$$\frac{\text{Basic salary} + \text{Dearness allowance} + \text{City allowance}}{150 \text{ hours (fixed)}}$$

$$\text{Suppose} = \frac{\text{Rs. } 500 + \text{Rs. } 200 + \text{Rs.} 10}{150} = \frac{710}{150} = \text{Rs. } 4.73$$

$$= \text{Rs. } 4.73 \text{ per hour.}$$

Thus, the authorities of the Government Press make payment for overtime work to the permanent staff on the basis of the above two systems. These two systems are applied for the calculation of payment of overtime works in different seasons depending upon the variations in the working hours of the press. The payment of overtime to daily wage workers is made on the following basis:

Daily Wage workers

$$\frac{\text{Wage fixed per day}}{\text{Total No. of hours of overtime}}$$

$$\text{Suppose} = \frac{\text{Rs. } 10}{4 \text{ hours}} = \text{Rs. } 2.50 \text{ per hour}$$

OR

Wage fixed per hour × Total No. of hours of overtime
Suppose, Rs. 2 × 4 hours = Rs. 8 per day.

The above two systems for calculation of payment of overtime work to daily workers are applied by the press authorities with the variations in the working hours of the press and urgency of works. In normal condition, the workers on daily basis receive their payment for extra work at the rate of Rs. 2 per hour per day on the working days. This category of workers is not allowed to work overtime during holidays including Sundays. The workers engaged for this purpose, however, are mostly sons and relatives of the permanent staff of the press.

Other Facilities : Regarding other facilities under service conditions granted to the workers, the position of the press is as follows :

The basis of promotion adopted by the press is test, seniority and efficiency. Grievances of the employees are channelised through two workers' unions, *viz.*, Bihar Government Printing Press Employees Association and Bihar Government Printing Press Employees Union. Both the unions are associated with the Federation of Bihar Secretariat Employees Association. Disputes between the management and workers are settled through legally constituted Works Committee, Grievance Committee and Industrial Committee. The press workers rarely go to court for the settlement of their disputes.

The workers enjoy the benefits of Contributory Provident Fund, Gratuity, Pension, Bonus and Workmen's Compensation. But the press does not provide the benefits of Employees State Insurance to its workers. The workers get a scale of pay with yearly increments. But the pay scale of the press employees is comparatively inferior to their counterparts in the Secretariat. *Ad hoc* payment in cash on certain occasions and loan for house construction purposes are also granted to the press employees. Thus, promotion has been used as a method of incentive plan in the Government Press which is a good practice.

Trade Unionism

We have mentioned earlier that the Bihar Government Printing Press employees are represented by two unions: (1) Bihar Government Press Employees Association, and (2) The Bihar Government Press Employees Union. We shall now study the objectives and functions of these unions separately with a view to examine their impact on industrial relations.

Bihar Government Press Employees Association

"At first the Association of Government Press employees (Gulzarbagh) was established in 1922 as the Bihar and Orissa Government Press Workers Union".[3] After separation of Orissa from Bihar the workers union was reformed as Bihar Government Press Employees Association in 1935. The Association was registered under the Indian Trade Union Act, 1926. It was neither affiliated with any political party nor with All India Trade Unions. The Association has been attached with one of the Federations of Bihar Government Employees. It has got the leadership of some celebrated labour leaders like Shri Mohit Mohan Ghose, Shri J.N. Chatterjee, Shri Rewati Mohan Das, Shri Jamunadhar Mishra, Shri Ganga Sharan Singh and Shri J.N. Chaudhary. At present the Association is working under the presidentship of "Tiger" Yogendra Narayan Singh, an employee of the Government Secretariat. Out of a total of 673 press workers, the Association has a membership of 238. The Association is financed by the membership fee of 50 Paise per month supported by occasional collections. The main purpose of the Association is to attend to the labour problems of Government Printing Press. The methods used for settlement of disputes by the Association include participation in Works Committees, Grievances Committees and Collective bargaining.

Bihar Government Press Employees Union

The Bihar Government Press Employees Union is another association for the press workers in the Government Press. The Union was established in 1977. It is duly registered under the Indian Trade Union Act. This Union too is a non-political organisation. The Union is associated with another faction of the Bihar Government Employees Federation led by Shri

Yogeshwar Gope. At present Shri Gope is the President of the Union. The total membership of the Union is about 600 which consists of 80 workers of Ranchi Government Press and 60 workers of Secretariat and Raj Bhawan Presses. The main activities of the Union relate to protection of workers' interest and organisation of demonstration or strikes. The membership fee is 50 Paise per month per worker. Occasional collection from the members is a supplementary source of finance to the Union.

Thus, we find that there are two parallel labour organisations in Government Printing Press. Both the Unions have their separate identities and they are affiliated to different employees federations.

Labour Movement

The Bihar Government Press Employees Association is the oldest association and it remained as the only union of press workers in the Government Press till 1976. The Association has got several successes and failures during a long history of its existence. In 1922 the management rejected outright the demands of the press workers. The demands were in connection with fixed time for the payment of wages, fixed working hours according to Factories Act, 1911, security of jobs and permanent nature of job. The workers went on strike and the management declared a lock-out in the press on 25th February, 1922. This strike was considered to be an important labour movement in the history of industrial disputes in India, because strike was a rare action during 1922. The leaders successfully continued the strike for three months. This shows the Unions' strength and unity. "During the strike period seven employees of the press were dismissed".[4]

In 1923, an enquiry into the above demands of the workers was made by Mr. Collin, Director, Department of Industries, Government of Bihar. A report was submitted in which all the demands of the workers were favourably considered. This award was implemented in 1924.

During the period of economic depression (1929-1939) frequent changes in the working shifts were made which caused a monetary loss to piece-rate workers. The Association prepared a memorandum of its demands regarding fixation of half holidays for piece-rate workers and an increased payment for their work. The memorandum was submitted to the Governor of Bihar. The demands were favourably considered and implemented.

In 1939, when Defence of India Rules were introduced, the press workers concentrated on production rather than Union activities. In 1947 the Association was reorganised and the workers took full interest in the Trade Union activities. Soon after its reorganisation they presented two demands to the management. The demands included fixed channels of promotion and revision of pay scale of the press employees. A Committee under the Chairmanship of Shri Anugrah Narayan Sinha, the then Finance Minister, Bihar was constituted in 1947. The Committee favourably considered the demands of the workers and recommended for the avenues of promotion of non-clerical staff (industrial staff) in detail in all sections, *i.e.*, Printing, Composing, Binding and Reading Branch. The pay scale of the workers was also revised.

In 1948, the Bihar Government Press Employees Association placed a list of its demands before the management. The demands were in connection with the formation of Works Committee, appointment of a Labour Officer in the press, schedule for the opening and closing of main gate, provision of a canteen, housing facility, payment of salary and wages in time, rest house, separate arrangement of hospital for the Government Press employees, cycle stand, presence of officers in time, facilities of first-aid, accident compensation to workers, adjustment of casual leave, medical leave, and earned leave against their salary during absence from duty, timely confirmation of the employees and other problems. For the fulfilment of these demands the press workers demonstrated before the press authorities on several occasions. The workers submitted the list of demands to the then Chief Minister of Bihar also. The Chief Minister in consultation with the then Labour Minister took a favourable decision and some of the demands were met after 1954, particularly those relating to formation of a Works Committee, appointment of a Labour Officer and provision of a canteen.

From 1958 to 1975 the press workers continued their struggle for other demands which had not been fulfilled earlier. In course of time some new demands were also included in the list, *e.g.*, revision of pay scale, payment to daily workers on the basis of work done and appointment of a Printing Engineer in the press.

In 1977, a rift arose among the leaders of the Bihar Government Printing Press Employees Association for leadership with the result that a separate Bihar Government Press Employees Union was formed in the same year. Although both the Unions of press workers have their separate identities having inside leaderships, non-political organisation and affiliation to separate factions of Bihar State Government Employees Association, their objectives and function remained the same. Both Unions organise meetings separately, their charter of demands more or less resemble and they often demonstrate before the press authorities and the Government of Bihar jointly. Truly speaking, the nature of work of the employees of the Government Press is different from that of the employees of Bihar Government offices. Moreover, the demands of the press employees are sometimes of a technical nature. The pay scale of the press employees is far lower than that of the employees of Government Offices. Yet the press Unions are fully associated with Bihar State Government Employees Association. In 1982, the pay scale of Bihar State Government Employees was revised. The press employees have also been granted a new scale of pay but not at par with their counterparts in the offices of the Bihar Government.

Major Problems

Our discussion with a number of Sectionholders in printing, composing and binding sections and with the leaders of both the Unions, reveals that the problems of the Government Press are different. The State Government is the main employer of the press workers, but it is not their supervisor. Most of the grievances of the press are technical in nature; and Government is guided by the opinion of the experts unknown to the printing technology. In this situation it is difficult for the press employees to convince an officer of the press and the government about their operational problems. Instead of examining the real state of affairs relating to a problem, workers often receive transfers, dismissals, or bad records. On several occasions the press authorities have submitted misleading reports and distorted facts about the workers demands to the Government. The peaceful efforts have failed to redress the grievances of the employees.

The appointment of a Printing Engineer has long been pending in the press. The press Unions have also been demanding for this appointment. They have also been suggesting that qualified foreman should be recruited for printing, composing and binding sections of the press. But none of the Press Superintendents since 1921 have shown any initiative to improve the quality of the organisation. Sometimes appointment of efficient technical staff is discouraged because the senior-most executive himself is not a technical expert. This set-up has its other implications too. Required materials in sufficient quantity are not supplied to the workers. This badly affects the quality of printing and composing jobs. Hence, there is the need for greater awareness on the part of those who are responsible for the immediate management of the Government Press to show their talent in promoting production efficiency and labour satisfaction.

REFERENCES

1. Government of Bihar, *Bihar District Gazetteers, Patna,* Gazetteers Branch, Revenue Department, Patna, 1970, p. 175.
2. Government of Bihar, *Secretariat Instructions,* Revised Ed., 1952, pp. 167-169.
3. Bihar Government Press Employees Association, *Review of the Working of the Government Printing Press Employees Association,* 1946-47, p. 3.
4. *Ibid.*, p. 5.

6

University Press

THE purpose of this chapter is to examine the organisation and operation of one of the oldest presses in Patna, *i.e.*, Patna University Press. This press is fully owned by the University and it caters to the printing needs of Patna University.

Patna University Press was established in the year 1951 to meet the printing requirements of the University. The press is registered under the Factories Act, 1948. It was started at first in a hired building at a monthly rent of Rs. 400 at Dharahara Kothi, Naya Tola, Patna providing an area of about twenty-six hundred square feet. In the year 1980, however, the Press was shifted to its own building, east of Moinul Haque Stadium, Rajendra Nagar, Patna. The Press now covers an area of thirteen thousand six hundred square feet.

Organisation Structure

Nature of Organisation

Patna University Press is a University Press; manned both by office staff and workshop staff. Since it is a part of the University, Registrar is the administrative head to control its operations. But with a view to streamlining its working, a Press Management Committee was constituted by the University authorities in 1967. It was mainly to look after the day-to-day affairs of the Press. The Press Management Committee was constituted in accordance with the recommendations made by the University Enquiry Committee appointed in 1966.

Enquiry Committee

Though the Patna University Press was established in 1951, its management often received comments and criticism by the administrators of the University. The problem of efficient management of the press, therefore, remained a topic of continuous discussion. It was in the light of these growing criticisms that the Vice-Chancellor appointed a Press Enquiry Committee on the 11th April, 1966. The terms of reference of the Committee included examination of its accounts, overtime payments, paper purchases, hours of work, income utilization and stores verification. The Committee made the following recommendations:

(1) a thorough overhauling in its organisation and administrative procedures;
(2) creation of a Standing Committee to dispose of all matters relating to the press;
(3) complete accounting of press transactions at the press itself;
(4) improvement in press planning through balancing of work, determining of loads and definition of efficiency; and
(5) development of composing section so as to feed the printing capacity of the press at a satisfactory level.

The main recommendations of the Press Enquiry Committee received the approval of the University Syndicate which resolved in November 1967 to constitute a separate Management Committee to look after its day-to-day affairs. It consisted of a Chairman, an Executive Officer and four other members. The Press Management Committee was conceived as the chief executive body for the press responsible for its effective working.

The Press Management Committee made various investigations into the deficiencies of the University Press and tried to improve its performance. In November 1968 a sub-committee on press reorganisation was appointed to suggest appropriate measures. The sub-committee made recommendations regarding pay of the employees, organisation structure, technical requirements and standing order. It was in the light of these recommendations that the University Press was subsequently reorganised and a methodology of work was developed.

Procedure of Recording Work

The University Press maintains various records for the job done. Some of these records are:

(*i*) Job register;
(*ii*) Bill register;
(*iii*) Job docket;
(*iv*) Demand register;
(*v*) Challan register;
(*vi*) Ledger book;
(*vii*) Cash register;
(*viii*) Work Process register; and
(*ix*) Printing Press register.

The position of the University Press in the procedure of recording work may be compared to other presses of the town in the following manner (Table 6.1):

TABLE 6.1 : **Procedure of Recording Work**

System Record	No. of Presses maintaining the record	Percentage to total
(*a*) Job Register	56	41.5
(*b*) Docket Register	3	2.2
(*c*) Demand Register	1	0.7
(*d*) Work Cards	2	1.5
(*e*) Challan Register	4	2.9
(*f*) Bill Register	23	17.1
(*g*) Cash Memo	7	5.2
(*h*) Other (order book)	7	5.2
(*i*) No such system	32	23.7
Total	135	100.0

Note: Some of the presses follow more than one system of recording work.

Table 6.1 shows the various types of records maintained by the presses of Patna and their percentage distribution. The table shows that job register is maintained by almost half of the presses. This, however, is the most common type of record maintained in the printing industry in Patna. The other records are not so frequently used by the printers. In this respect, therefore, the University Press is in a slightly better position.

Financial Management

We shall now examine the financial aspects of the University Press. The press is entirely dependent upon the University for its financial needs. Therefore, the financial operation is guided and controlled by the University authorities.

Ba is of Payment

Payment to the staff is made by the Press on the following basis:

	Section	*Basis*
1.	Printing	Salary
2.	Composing	Salary and contract
3.	Binding	Salary
4.	Proof Reading	Salary

Budget and Accounts

Proper accounting system and timely compilation of figures help management in quicker decisions. Our enquiry regarding the Profit and Loss Account and Balance Sheet of the Press shows that the University Press does not prepare such statements of accounts. However, an attempt was made in this direction after 1968. But this too could not succeed for one reason or the other. It may be relevant to suggest here that the University Press should prepare its periodical Profit and Loss account and Balance Sheet as these two statements are the basis of an effective budgetary system and accounting particularly in a developing public enterprise. Balance Sheet is a statement which shows the position of the assets and liabilities at a glance at a particular point of time. One can avoid preparation of this document only at the peril of being duped and deceived in respect of the real state of affairs of a particular organisation. Hence, we suggest that annual Balance Sheets must be prepared for the University Press as a part of its accounting system.

"It may, however, be pointed out that these two documents

can only be prepared when proper stock-taking is done in respect of :

(*i*) paper;
(*ii*) work in progress;
(*iii*) ink;
(*iv*) metals, etc".[1]

Size of Out-turn

An important area of our investigation has been capacity evaluation of the University Press. Such an evaluation, however, is a difficult task. Any such attempt has to be based on a number of assumptions. An estimate of the production capacity of the University Press may be made in the following manner.

Estimated Production Capacity

1. Effective working hours			
(*i*) 1st July to 31st March	—	—	6 hours a day
(*ii*) 1st April to 30th June	—	—	5 hours a day
2. Total number of holidays	—	—	100
3. Total number of working days @ 6 hours per day		190×6=1140	
4. Total number of working days @ 5 hours per day		75×5= 375	
Total		1515	

Less provision for :	
(*a*) Casual leave	16
(*b*) Earned leave	30
Total	46
Less Sundays	6
	40

5. Total hours of deduction (–) 40×6=	240
6. Effective No. of working hours per year	1275

7. Printing capacity per hour for
   1000 impressions:

Two flat machines	= 3 Formes
Two Treadle machines	= ½ Formes
	3½ Formes

Calculating the charges for double demy forme of 16 pages at the most moderate rate of Rs. 60 per forme, the total amount comes to $(7/2 \times 60 \times 1275) =$ Rs. 2,67,750. This out-turn, however, does not include many charges, *viz.*, cost of paper used, binding, stitching, cutting, perforating, die-printing, etc., which would easily take the annual out-turn to over Rs. 3,00,000. As against this extremely conservative estimate the annual average out-turn of the press for five years (1960-61 to 1964-65) works out to Rs. 1,06,000 only. And even today the annual out-turn of the University Press has not improved much. This is apparent from Table 6.2.

TABLE 6.2 : **Size of Out-turn**

Year	Annual out-turn (Rs.)
1965-66	61,723=44
1967-68	1,03,348=17
1970-71	1,57,421=42
1974-75	1,81,918=35
1975-76	1,98,796=66
1976-77	1,39,159=14
Total (6 years)	8,42,367=18

Table 6.2 shows that during the last six years within ten years, the total amount of work done by the press was Rs. 8.4 lakh which gives an average of about Rs. 1.4 lakh per annum.

After 1977, the system has completely changed. There is complete absence of any organised system of accounting. Profit and Loss Account or Balance Sheet is not prepared. It is, therefore, difficult for anybody to know the annual out-turn of the press since 1977.

Yet Table 6.2 further shows the gap between the estimated

production capacity and the actual out-turn of the press. This should be a matter of concern to the appropriate authorities and steps should be taken to bridge this gap as quickly as possible.

The valuation of the out-turn is done on the basis of a schedule of rates approved by the University in 1980. It is necessary to revise the old schedule on more scientific lines and on the basis of the costing done by the Mudrak Parishad (Bihar). Rates charged outside Patna seem to have little relevance for a press situated at Patna and affected by the wages and other rates obtaining here. Hence, the need for an immediate revision of this schedule taking into account the current inflationary changes.

There is another problem. Individual out-turn records appear to be slightly defective. They do not in all cases mention the specific work done. They are merely quality records showing the number of pages composed or galleys corrected or formes printed, not the page numbers and the specific work to which they relate. This practice leaves opportunities for preparing duplicate bills for the same work and permits inflation of out-turn still further. It is also necessary to plug this loophole to avoid manipulation of accounts.

Presentation of Budgets and Accounts

The question of a suitable method of presentation of accounts is connected with some other issues of wider importance. Run on the departmental basis, the present system of budgeting and accounting of the press is not satisfactory. There is no effort to estimate the amount of profit earned or loss sustained in the working of the press. This appears to be thoroughly inconsistent with the nature of the work involved. The press, irrespective of whether or not accepts outside work, must be run on business principles. For it is difficult to understand why all the botherations of owning and running a press should be accepted if it does not make a surplus of return for the University, or at least does not involve it in a loss. "Any calculation of the result of the working of the press is rendered difficult under the present method of presenting its accounts".[2] Firstly, the system of presentation is wrong; and secondly, it fails on many occasions. Some of the major weaknesses in the accounting system of the press are—no profit no loss account is prepared, the classification of items is incomprehensible, depreciation on machineries is not charged, provision for replacements is not made, system of reserves does not exist, cost of accounting department of the Press is not debited to press accounts, etc.

It was in view of these weaknesses that the Patna University Press Enquiry Committee (1966) recommended for an immediate change in the system of presentation of accounts of the press. Subsequent to this recommendation the University did make some changes in the presentation of accounts of the press. Yet much more substantial changes in the preparation of Profit and Loss Account and Balance Sheet are needed. The suggested forms appear in (Appendix XA and B).

Technical Position

Machine

The number and types of machines used by the University Press are as follows:

Machine	*Number*
(*a*) Flat	2
(*b*) Treadle	2
(*c*) Ruling	1
(*d*) Stitching	1
(*e*) Cutting	1
(*f*) Mono Operating	2
(*g*) Mono Casting	2
(*h*) Die Stamping	1
(*i*) Others; Perforating, Numbering, etc.	6

The press mostly uses two systems of composing—hand composing and mono composing, the latter being a system of mechanical type setting. On personal observation it has been found that hand composing is adopted in most of the jobs because of old and obsolete mono casting machines. Other systems such as Inter-type Ludlow and Photo-composing are not used by the press.

Materials

For paper, metals and chemicals the press mostly depends on local market. Purchase of these materials is made only when the requisite authorisation from the University is obtained. Types are melted and manufactured by its own Mono Casting machines.

Labour Conditions

The press is run with the help of the following categories of staff :

Staff

Workshop	*Number*	
(*i*) Compositors	11	
(*ii*) Machinemen	5	
(*iii*) Binders	3	
(*iv*) Coolies	4	
(*v*) Computer	1	
(*vi*) Proof Reader	1	
(*vii*) Mono Operator	1	
(*viii*) Mono Caster	1	
(*ix*) Darwan	2	
Total		29

Office

(*i*) Superintendent	1	
(*ii*) Accountant	1	
(*iii*) Assistant	2	
(*iv*) Typist	1	
(*v*) Peon	1	
Total		6
Grand Total		35

This shows the number and categories of workshop and office staff who serve the University Press. Obviously, the technical staff is almost 83 per cent of the total staff.

Working Conditions

The press has made some arrangements for health, safety, cleanliness and welfare facilities. The staff employed by the Press are of three categories—permanent, temporary and contract. The number of employees in these categories is 29, 2 and 4, respectively. Leave rules are the same as applicable under the University statutes. In other words, the press staff too are entitled to get earned leave, casual leave, medical leave and extraordinary leave. This is besides the usual holidays granted to the University office staff. House Rent Allowance is granted in accordance with the University rules. In case of penalties for offences warning is the usual system. The press works overtime on certain occasions for about 3 hours per day. Payment for overtime is linked with the basic salary of employees, at the rate of 4 hours of overtime equal to a full days duty. The basis of promotion followed by the press is both seniority and efficiency. Grievances of the employees are channelised through the University Employees Association. The workers enjoy the benefits of Contributory Provident Fund and Employees State Insurance. Sometimes promotion is also followed as an incentive plan.

Trade Policy

Major Problems

Despite the support from University and being a part of its structure, the University Press has long been a problem organisation. The major problems facing the Press which have been identified in course of our investigation are as follows:

1. How to feed the Press?
2. How to improve the quality of work?
3. How to initiate quickness of operations?
4. How to help the financial position?
5. How to establish sound staff relations?
6. How to improve the placement?
7. How to improve the University policy towards the Press?
8. How to make timely allocation of jobs to the Press?
9. How to provide adequate accommodation to the Press?
10. How to reduce the cost of operations?

11. How to obtain efficient and effective utilization of capacity?
12. How to remove administrative dislocation?
13. How to give job satisfaction to the staff?
14. How to improve profitability and self-reliance of the Press?

These are some of those problems of the Patna University Press to which we shall now devote our attention with a view to offering suggestions.

Suggestions

On the basis of our detailed enquiry into the working of the University Press in particular and the printing industry in Patna in general, we may offer the following suggestions.

Output

Limited sources of work and poor supply of materials reduce the volume of output in the University Press. Efforts, therefore, should be made firstly, to increase the number of customers; and secondly, to secure more printing orders from the customers. It has been noticed that even some of the University departments or the constituent colleges do not use the University Press for their full printing requirements. Besides University work, the Press should also handle outside job, particularly of Government departments, statutory bodies, boards, corporations, etc., to augment its revenue. Accordingly the University may issue instructions to its various offices, departments and colleges to make full use of the Press by patronising it as much as possible for the printing job. Similarly, the Government Press may be approached to issue instructions to all government or semi-government departments in Patna to favour the University Press for printing assignments at the prescribed government rates. The Press Enquiry Committee too had considered this point and observed, "It is a fact that Patna University Press is not working to full capacity. If the University cannot provide work to keep the Press engaged to capacity, there is no reason why job work should not be undertaken to put the surplus capacity of the Press to profitable use. There is no definite or specific procedure prescribed for this purpose. It is, therefore, suggested that in the light of the capacity of the Press and the volume of work which the University itself can provide to the Press, the University should lay down a specific procedure for undertaking job work from outside agencies".[3]

Quality

It is often found that the quality and standard of printing done by the Press is not as good as it should be. It is partly due to the poor quality of materials available to the Press, but more due to the absence of proper attention towards perfection. Greater consciousness, therefore, is required among the Press staff to appreciate the value of quality and correctness in printing industry.

Promptness

Quickness of operations and promptness of delivery attract customer's satisfaction. But the University Press has not been able to develop adequate goodwill in these directions. There should be a systematic job allocation, an adequate policy of work and a proper follow-up measure. If sometimes urgent works have to be accommodated, they should be especially charged for the purpose. Otherwise a sequence of production should normally be followed.

Staff Discipline

The management has to devote some more attention to staff discipline in the Press. This is closely associated with the efficiency of the workers. If the Press can earn profits through efficiency it may be able to pay some bonus to its staff as well.

Efficiency

The University should send printing jobs to the Press well in advance. Unusual haste creates a lot of dislocation in the Press. This is necessary to develop efficiency.

Cost Control

Some attention should also be paid to examine the cost structure of the University Press *vis-a-vis* other private and public presses in the town. This may provide opportunities for reduction in the cost of production and improvement in the quality of operations.

Repairs and Replacements

Timely repairs and replacements are necessary for the operational efficiency of plant and equipment. Machines like mono casting, mono operating, die stamping, stitching, etc., remain idle due to some reason or the other. They should be brought to working order so as to facilitate fuller utilization of the potential capacity.

Hours of Work

The Factories Act permits a normal working day of 9 hours with an hour's break. The private presses in Patna work according to the Factories Act. The working hours of private presses in Patna are generally from 9 A.M. to 6 P.M. with an hour's break from 1 P.M. to 2 P.M. This gives them 8 effective working hours per day. They work according to this schedule throughout the year. They do not have any changes in the above working hours during the summer except the Government Press, which changes over to morning hours during the summer. The effective hours of work in the University Press, however, vary from one season to the other as is obvious from Table 6.3.

TABLE 6.3 : **Working Hours of Patna University Press**

	Actual hours per day
April to June	
Workshop staff—6 A.M. to 12 A.M. without rest	6
Office staff—10.30 A.M. to 5 P.M. with ½ hour recess	6
July to March	
Workshop staff—9 A.M. to 5 P.M.—with ½ hour recess	7½
Office staff—10.30 A.M. to 5 P.M.—with ½ hour recess	6

The following comments may be offered in respect of this system :

(*a*) The normal working hours of the University Press as compared with the working hours of standard private presses in Patna are short by 2 hours during the summer (April to June) and by ½ hour during the rest of the year (July to March).

(*b*) The number of holidays excluding Sundays in the University is about 52 every year. In case of other presses, however, the number of holidays is restricted to 23.

Table 6.4 further indicates how the staff of the University Press have shorter working hours in comparison to the staff of other standard presses in Patna.

TABLE 6.4 : **Effective Working Hours**

Name of Press	Months	No. of days	No. of holidays	Holidays Sundays	Total effective days	Effective hours per day	Total effective hrs./year	Difference of hrs.
Other Presses	January to December	365	23	52	290	8	2320	
Patna University Press	April to June	91	6	13	72×6=432			
	July to March	274	46	39	189×7½= 1418		1850	470

As per Table 6.4, the effective hours of work of the University Press staff are about 470 hours less than other presses in Patna. It is, therefore, suggested that the working hours of the Press should conform to the normal working hours of other presses so as to get a better utilisation of its resources.

Cash Book

Before 1966, the Receipt and Payments Account of the University Press was maintained in a separate cash book at the University office. But the cash book was not checked regularly by any superior officer of the University. After 1966, however, a complete recording is made of all incomes and expenses of the Press in a separate cash book maintained at the Press itself. Moreover the cash book is regularly checked by the government auditors.

Outstanding Bills

The University office does not inform the Press as to which of its bills remain outstanding. It is then difficult for the Press to make efforts for the collection of the outstanding bills. Thus, a unified system of accounting calls for an immediate unification of the two sections—the University office and the University Press.

Technical Skill

The life of the machines is much influenced by the manner in which they are handled. But the handling of machines depends on the technical competence and skill of the personnel. This also depends on the attitude and care of the authorities in this area.

Our enquiry into the actual condition of the machines in the University Press shows that one mono machine is junk and completely out of order for the last 20 years. Even the other machines are not in as good condition as they ought to be. This type of apathy towards machine maintenance brings down the quality, volume and speed of output, increases the number of breakdowns and enhances the consequent expenses to bring them in order. The University is often requested by the Press to provide additional funds for machine repairs. But despite repeated requests this has not been done so far. We, therefore, suggest that repairing of idle machines should be taken up immediately and necessary provisions should be made in the budget for repairs and maintenance every year.

Depreciation

The University Press is running with the help of costly machines. According to sound accounting principles, the total cost of the machinery has to be charged to revenue over the operative life of the asset by way of annual depreciation. This should be incorporated in the accounting policy of the Press at the earliest.

Resources

The Patna University has acute shortage of funds. This hampers the day-to-day management of the Press as well. The Press has no capital of its own. Only the revenue finances its operations. Whenever bills remain unpaid, therefore, the Press is in a tight financial position.

Besides, the Press is entirely dependent on the University even for its petty expenses. The Press has a separate bank account, but transfer of money from University account to Press account every now and then is a difficult process. To be effective, a Press should be a self-contained unit with adequate reserves. This may be in the form of an advance from the University which can also help abnormal repairs, bad debts, contingent losses, etc. To further improve its financial position, the University Press may be equipped to handle the printing of confidential papers including examination, jobs of outside institutions. This may be a lucrative source of revenue to the Press.

Overtime and Contract

The University Press sometimes gets its work done on overtime basis or on contract. According to Factories Act the overtime work has to be paid for at double the ordinary rates. It is primarily for this reason that work done outside normal working hours is mostly done on contract basis. The general concept of contract work is that payment would be made on the basis of the work done. Payment is also sometimes calculated on the basis of time spent in completing a job. Time rate and piece rate for such work outside normal working hours have been approved by the University and payment is made on the basis of these approved rates.

The need for such work arises particularly at a time when urgent work is received and the Press has to work beyond its normal capacity. The urgency of the work necessitates the use of contract system as the only method of completing the work. Since such pressure of work in the University is a normal feature in November and March. These two months are the busiest season for contract work in the Press. The Press Budget, however, does not show separately payments for overtime or contract work. The words used in the budget for this nature of payment is 'payments to contingent staff'.

The contract rates once approved cannot be used for all time to come. They should be revised periodically. There is need for revising these rates to remove inefficiency and slow work. The rates may finally be approved by the University authorities.

These are a few suggestions for improvement in the working of the Patna University Press. All such attempts have to primarily aim at financial self-reliance and management development. No doubt the Press has shown some improvement after the appointment of the Press Management Committee. But it is only a joint effort of the University administration, Press Management Committee, Press Supervisor and the Press workers that can turn this unit into a sound and efficient printing centre.

This much about the Patna University Press. We shall now examine the working of the Bihar State Text Book Press in the chapter which follows.

REFERENCES

1. *Report of the University Press Committee constituted by the Vice-Chancellor to examine the Working of the Patna University Press*, 1966, p. 9.
2. *Ibid.*, p. 8.
3. *Ibid.*, pp. 18-19.

7

Text Book Press

General

"BIHAR State Text Book Publishing Corporation Limited, a Government of Bihar undertaking, was registered under the Companies Act in September 1965. While setting up this Corporation, State Government, in consultation with Planning Commission, decided that it should have fully equipped modern printing press of its own".[1]

The need for the establishment of a Text Book Press in Bihar was felt in 1965. Accordingly the Bihar State Text Book Publishing Corporation Ltd., was formed for the purpose of printing and publication of text books up to Secondary level in the State of Bihar. Though the idea to establish a press mooted in 1965, it took almost seven years to implement the decision.

Brief History

Mr. S.N. Sinha, the then Education Minister, Government of Bihar, and Chairman of the Corporation asked the former Managing Director of Bihar State Text Book Publishing Corporation Ltd., Late Mr. M.P.N. Sharma to take early steps to set up a modern printing press at the earliest opportunity. It was Mr. Sharma who took initiative in the matter and discussed the proposal with the Adviser and Joint Adviser, Ministry of Education, Chief Controller of Printing, Government of India and the Technical Experts of the Maharashtra and Andhra Pradesh Text Book Presses. The problem of hard currency was solved through assistance from East Germany Printing Plant. In

1965, Indian Parliamentary delegates visited the printing plant of East Germany to make contract for the project. Dr. Sperling, Deputy Director, Polygraph-Expert of East Germany visited Patna in 1966. A formal discussion with Dr. Sperling at the State level was held and it was agreed in principle to import printing machines from Rupee Block Area on tender basis. Then the necessary premission was obtained from the office of the Chief Controller of Imports, Government of India. A work analysis was prepared and a tender for a composite plant on turn-key basis was issued. Out of the different tenders received, the offer of M/s Polygraph-Expert of G.D.R. was accepted on December 2, 1967.

Plant Design

The design for the Bihar Text Book Press was prepared by the Central Project Bureau of Polygraph at Leipzig. The design is based on the latest technical innovation in order to ensure that material, usage, labour and energy consumption are kept to a minimum. The different components of the plant were designed in such a manner as to give maximum return within minimum period. The project, therefore, was fully conceived for sophistication in both design and technology. Necessary clearance of the project was obtained from the Director General of Technical Development, Chief Controller of Printing and Ministry of Finance, Government of India. Official delegates of the Government of Bihar were sent to Berlin twice in 1969 with a view to finalising the project. All this resulted in the signing of two contracts for the project at Berlin, one for the supply of machineries and the other for manpower training.

Cost of Project

The Government of Bihar sanctioned a sum of Rs. 1.22 crores to the BSTBPC in 1969 for the establishment of the Press. The cost of the machine components of the project was Rs. 62 lakhs out of which a sum of Rs. 20 lakhs was paid as advance to Polygraph-Expert in September 1970. For the balance amount of Rs. 42 lakhs, guarantee was furnished by the Central Bank of India on the basis of the counter-guarantee of the State Government. This amount was to be paid in course of 8 years on deferred credit with an interest of 4½ per cent per annum on the outstanding balance.

Installation

The first shipment of the machines started in July 1971. The rest of the machines were received by February 1972. A batch of G.D.R. Engineers headed by Mr. Otto Penndrof, Chief Engineer, arrived in Patna in March 1972. The installation of machines was taken up on 15th March, 1972 and one unit of the Press started production on 29th March, 1972. By the end of 1972, however, all machines were installed and the Press was fully ready to print books for the Government. The production started and in 1972 itself the value of the output of the Press was to the tune of Rs. 38 lakhs.

Thus, with an active and sustained interest of the State Government and a strong link in the chain of Indo-G.D.R. Economic Cooperation, the Bihar Text Book Press was established. The Press is registered under the Factories Act, 1948. It works in its own building with an area of about 54,400 sq. ft. situated at White House, Budh Marg, Patna-1. After all the installations were completed, this came out as one of the most modernised presses in the country having 75 machines including a complete set of automatic binding machines.

Organisation Structure

Nature of Organisation

The BSTBPC was incorporated in 1965 under the Companies Act, 1956 primarily to take over the Government business relating to text books and to publish, print, sell, supply or otherwise deal in text and other books and literature as may be prescribed and/or approved or required by the Government of Bihar. Thus, the Corporation is a Government of Bihar undertaking. The management of the Corporation and its Press is entrusted to a Board of Directors whose membership is determined by the Governor from time to time within the prescribed minimum and maximum limits under its Articles of Association. The Board is composed of Directors appointed by the Governor as well as Directors nominated by such institutions, bodies and authorities whom the Governor may direct or authorise for the purpose. No term is prescribed for the Directors. They hold their office for such period as they are appointed for or until they are removed or replaced or otherwise cease to hold the office. "The Governor also appoints the Chairman and the Managing Director and can confer upon them such of the powers exercisable under the Articles by the Directors as he may think fit and can also, from time to time, revoke, withdraw, alter or vary all or any of such powers".[2] The powers to appoint important officers of the Corporation like Financial Adviser and Chief Accounts Officer, etc., also vest in the Governor. "The Board of Directors manage the business of the Corporation subject to the provisions of the Act and the instructions or directions of the Governor. However, the power to issue directives or instructions extends not only to matters relating to policy but also to the finances and conduct of business and affairs of the Corporation or Directors thereof".[3] Certain powers are exercisable by the Board only with the approval of the Governor, *e.g.*, the power to raise or secure payment or repayment of any sum or sums of money. Some matters are specially reserved for decision of the Governor by the Articles. They relate to any programme of capital expenditure exceeding rupees one lakh, creation of and appointment to all posts whose maximum salary exceeds Rs. 800; granting of a loan or the giving of a guarantee or any other financial assistance to any one particular concern exceeding Rs. 1,00,000, etc. Even the Revenue Budget of the Corporation is subject to the approval of the Governor [Art. 94(1)].

At the end of 1983, the Board of Directors of BSTBPC consisted of 13 members including the Chairman and the Managing Director. The wide ranging powers of the Governor (or the Government) together with the composition of the Board which included the Minister of Education (as Chairman) besides other officials of the Departments of Education and Finance made BSTBPC functions more like a Government department rather than an autonomous commercial enterprise. Many policy matters of financial and operational nature in respect of which either the approval of the Government was required or the decision was reserved for the Government restricted its smooth operation. Delayed decisions often hamper efficiency of commercial operations. Moreover, the undefined tenure of Chairman or members of the Board, coupled with the deputation of serving government officers as chief executive, results in frequent changes in top position adversely affecting the continuity in policy and direction.

The management and organisation of the Corporation consist of the following organisation structure:

Governor
Board of Directors
Education Commissioner

Office	1. Managing Director
	2. Financial Adviser
	3. Secretary-cum-Accounts Officer
	4. Personnel Officer
	5. Section Officers
	6. Assistants
Academic Section	1. Academic Registrar
	2. Production Officer
	3. Senior Proof Reader
	4. Artist
	5. Proof Readers
Press	1. Works Manager
	2. Production Officer
Printing Section	1. Technical Officer
	2. Operators
	3. Operatives (Machinemen)
	4. Shramik
Composing (Hand & Mono) Section	1. Technical Officer
	2. Operatives
	3. Shramik
Binding	1. Technical Officer
	2. Assistant Technical Officer
	3. Operatives

Power Section	1. Technical Officer (Engineer) 2. Operatives
Sales Section (Finished Store)	1. Sales Officer 2. Sales Manager (Head Office) 3. Sales Manager (Hazaribagh, Muzaffarpur, Bhagalpur) 4. Assistants
Store (Materials)	1. Works Manager
Security	1. Security Officer 2. Security Guard

Procedure of Recording Works

The Text Book Press maintains various records for its different jobs. Some of the important records are:

(*i*) Job distribution register (job ticket);
(*ii*) Job docket;
(*iii*) Order book;
(*iv*) Job receipt register;
(*v*) Challan register;
(*vi*) Bill demand register;
(*vii*) Cash book;
(*viii*) Sales register; and
(*ix*) Estimate Sheet (Individual performance).

The Press maintains almost all essential types of registers. The individual performance of the workers of the two sections, *i.e.*, composing and printing, is maintained by the Works Manager through manual computing system. Jobs received from outside sources are also duly recorded in a separate register. Moreover, separate registers are maintained for all sections of the Press.

Financial Management

Capital Structure

BSTBPC was registered as a wholly-owned Government Company in April, 1965. Its major purpose was to provide an efficient system of production and distribution of text books in all subjects for schools within the State. It was conceived as a printing and publishing corporation with an authorised capital of Rs. 1.50 crores divided into 15,000 equity shares of Rs. 1,000 each. However, the paid-up capital of the Company is Rs. 11.25 lakhs only and has stayed at that figure since inception.

When BSTBPC assumed the duties of the erstwhile Text Book Committee in April 1966, the value of the assets transferred to it estimated at Rs. 7.96 lakhs. This was treated as loan to the Corporation carrying interest at the rate of 10 per cent per annum. Advances were made to BSTBPC from time to time for construction of Press and office building, purchase of printing machines and for working capital. All these advances were in the form of loans with the result that soon the Corporation's capital structure became highly leveraged with a heavy weight of fixed charges.

Table 7.1 shows the changes in the capital structure of BSTBPC between 1970-71 and 1976-77.

It is clear from Table 7.1 that the equity capital of BSTBPC declined in amount and relative proportion between 1970-71 and 1976-77, while the long-term debts registered sharp increase during the same period. The paid-up capital of BSTBPC remained stationary at Rs. 11.25 lakhs during the period, while Reserves and Surplus declined from Rs. 13.58 lakhs in 1970-71 to Rs 12.64 lakhs in 1976-77. "It is doubtful if BSTBPC had any real Reserves and Surplus after 1971-72 since losses amounted to Rs. 34.52 lakhs in 1972-73".[4] Long-term loans which amounted to Rs. 56.18 lakhs in 1970-71 accounting for about 42 per cent of the total liabilities in that year increased to Rs. 101.18 lakhs, *i.e.*, around 56 per cent of the total liabilities in 1974-75. The amount of such loans came down to Rs. 95 lakhs in 1976-77 but its percentage went ahead (*i.e.*, 56%) because of fall in the total equity capital. Thus, as against zero growth in equity capital, long-term loans increased by about 80.1 per cent. The capital structure of BSTBPC had yet another feature of financing fixed assets like plant and machinery and building entirely from loan. In the case of private business, the Corporation usually raises sufficient share capital not only for the cost of fixed assets but also for the cost of establishing the business and also for permanent or regular

TABLE 7.1 : **Capital Structure of BSTBPC (1970-71 to 1976-77)**

(*Rs. in lakhs*)

Particulars	1970-71	1971-72	1972-73	1973-74	1974-75	1975-76	1976-77
Paid-up capital	11.25 8.42%	11.25 7.78%	11.25 6.11%	11.25 6.78%	11.25 6.23%	11.25 7.47%	11.25 6.65%
Reserve and surplus	13.58 10.16%	11.94 8.26%	12.55 6.82%	12.64 7.62%	12.64 7.00%	12.64 8.40%	12.64 7.47%
Government Loans	56.18 42.02%	79.18 54.75%	84.18 45.73%	91.18 54.98%	101.18 56.01%	101.18 67.19%	95.00 56.16%
Current Liabilities	52.65 39.40%	42.25 29.21%	76.09 41.34%	50.78 30.62%	55.57 30.76%	25.35 16.84%	50.28 29.72%
Total	133.66 (100)	144.62 (100)	184.07 (100)	165.85 (100)	180.64 (100)	150.62 (100)	169.17 (100)

Source : Data compiled from Annual Reports of BSTBPC for the relevant years.

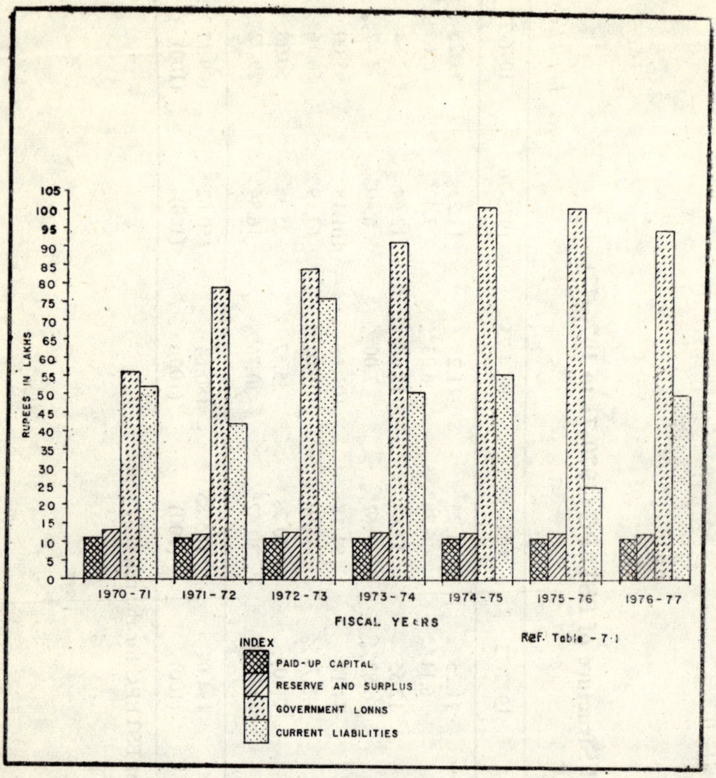

FIG. 7.1 : *Capital Structure of BSTBPC (1970-71 to 1976-77)*

working capital requirements. The paid-up capital of BSTBPC, therefore, appears to have been fixed without due regard to the actual requirements of the Corporation.

Basis of Payment

The Text Book Press makes payment to its staff on the following basis:

Section	*Basis*
(*a*) Printing	Salary and contract
(*b*) Composing	Salary and contract
(*c*) Binding	Salary, daily wage and contract
(*d*) Proof Reading	Salary and contract

Price Policy

The process of determination of cost and selling prices by the Corporation can be summarised as follows:

Paper cost per material	
(forme of 16 pages)	....
Labour cost	
(*a*) Establishment cost per forme	....
(*b*) Process cost per forme	....
(*c*) Printing cost per forme	....
(*d*) Binding cost per forme	....
Total	....
Material and labour cost per book	....
Total	....
Cost of storage, obsolescence,	
Depreciation, etc.	....
Total cost	....
Add 25% profit on cost	....
Selling price	....

Profitability

Table 7.2 shows the profits/losses of the Corporation during the last 17 years.

Table 7.2 shows that the profits of the Corporation have been erratic. The Text Book Corporation has two separate sections, *i.e.*, press and publication. A personal discussion of this author with the Secretary of the Corporation shows that the press section always runs in loss and it is the publication section which gives revenue to the Corporation.

From Table 7.2 it is also clear that the Corporation has sustained losses during 1971-72 and 1972-73. It was mainly because of the initial difficulties which a new unit of an industry faces. After 1973, however, the Corporation has regularly earned profit the amount of which varies from year to year. As a public enterprise of the State Government, therefore, the BSTBPC is one of the few profit-earning units.

TABLE 7.2 : **Profit and Loss of BSTBPC**

(*Rs in lakhs*)

Year	Net profit before taxes		Net profit after taxes
1966-67	6.24	Profit	0.01
1967-68	7.19	,,	0.01
1968-69	6.86	,,	0.03
1969-70	12.64	,,	0.01
1970-71	12.56	,,	0.00
1971-72	3.24	Loss	(Rs. 310)
1972-73	34.52	,,	...
1973-74	22.49	Profit	...
1974-75	20.99	,,	...
1975-76	34.29	,,	7.79
1976-77	120.11	,,	89.38
1977-78	76.03	,,	28.38
1978-79	95.94	,,	36.51
1979-80	129.53	,,	62.98
1980-81	65.89	,,	25.00
1981-82	90.37	,,	51.37
1982-83	44.50	,,	14.64

Source : Government of Bihar, Draft Annual Plan, 1984-85, p. 515.

Budgetary System

The Press prepares an annual budget through which it maintains its financial discipline. During 1984-85 the Corporation approved a sum of Rs. 8 crores as budgetary provision. The budget of the Corporation includes all receipts and expenditure in printing and publication of text books.

Size of Turn-Over

The size of turn-over of books of the Press has been fluctuating from year to year and it is much below the total annual capacity of one crore. This will be clear from the following descriptions. During the period of incorporation of the Press (*i.e.*, 1972-73) the total production of books was 38 lakhs only. It, however, increased to about 60 lakhs during 1973-74. At the end of 1977 the total print of books was 147.72 lakhs. out of which only 76.68 lakhs was printed in its own Press and the rest in the private presses inside and outside the State. Between 1978-80, the annual turn-out of books by its own Press varied from

38 lakhs to 70 lakhs, while the total prints have been in crores. This means that the fixed annual capacity of one crore of prints of the Press has never been achieved and on an average more than 50 per cent of the total production of books has been done outside the Corporation. The situation further deteriorates when the printing capacity and the actual production of books by the BSTBPC show a declining trend. This is mainly because of the frequent break-down of the printing machines due to inefficient staff, old and obsolete machines and negligence of the Press authorities, with the result that the average production of the Press reduced to 40 lakhs only during 1980-81, 1981-82 and 1982-83 as against the actual production of 3.50 crores prints, *i.e.*, on average production of 1.17 crores books. It means that the printing capacity of the Text Book Press has come down to a meagre 36 per cent. The rest of the 64 per cent printing work is done by other presses.

Technical Position

Machines

The number and types of machines used by the Press are as follows:

Name of Machines	*Number*	*Size (cm)*	*Speed per hour*
1. *Printing Section*			
(*a*) Treadle	1	13×19	1,500
(*b*) Flat Bed	1	61×86	1,500
(*c*) R.O. 62-Wave Offset	3	62	20,000
(*d*) Sheet Offset	3	70×100	4,000 to 12,000
(*e*) Auto letter printing	1		
	—— 9		
2. *Mono Section*			
(*a*) Mono Key-Board	3		
(*b*) Mono Casting	2		
	—— 5		

3. *Process Section*			
(*a*) Wheeler	1		
(*b*) Plate graning m/c	1		
(*c*) Drying m/c	1		
(*d*) Exhauster	1		
(*e*) Arc Lamp	1		
		5	
4. *Camera Section*			
(*a*) Automatic	1		
(*b*) Others	4		
		5	
5. *Cutting Section*			
(*a*) Automatic	3		10,000
(*b*) Hand	5		4,000
		8	
6. *Stitching Section*			
(*a*) Sewing machine	8		
(*b*) Flow line stitching	1		
(*c*) Threadless binding	1		1,000 to
		10	4,000 (Books)
7. *Others*			
(*a*) Litho	1		
(*b*) Folding	4		
(*c*) Gathering (Automatic)	1		
(*d*) Cover pasting	2		
(*e*) Pasting	4		
(*f*) Cover making	4		
(*g*) Drilling	1		
(*h*) Others	16		
		33	
Total		75	

The Text Book Press has four important departments: (*i*) Mono-type setting, (*ii*) Process (including Plate-making), (*iii*) Printing, and (*iv*) Binding. There are three Mono-Keyboards and two Casters. The main work in the type section is to prepare artpulls for the process department. The process department is equipped with latest photographic equipment all with electronic devices. The plate-making section is also equipped with electronic devices.

The printing section of the Press has three web offset machines and three sheet-fed machines. The special feature of web offset machine is that each machine can print four-colour at a time. Its production capacity is 20,000 formes per hour. The web offset machines are mainly meant for printing of text. Out of three sheet-fed machines two are of double colour each and the third one is of single colour. Even the sheet-fed machines can print 12,000 of 70 cm × 100 cm size sheets per hour.

The Binding Department has almost electrically operated machines of different types for different jobs. The important ones are cutting, gathering, threadless book binding and feeder wire stitcher machines. It is more due to the help of these machines in the Binding Department that it is possible to ensure timely delivery of printed materials in book form.

It is to the credit of the above-mentioned machines that even with a limited number of 17 operators and apprentices the Press continues to ensure its scheduled production.

Thus the Press has almost all types of latest equipment for composing, printing and binding works. Most of the equipments used by the Press are automatic. The Press uses both hand and mechanical devices for composing. Mechanical type setting comprises of mono, lino, intertype and photo composing systems.

Training

According to the contract approved by the Government of India, one Chief Engineer, eight engineers and four instructors were deputed by the Government of East Germany to the Bihar State Text Book Press for installation of machines and training of workers. All engineers left India after the installation of the Press but the instructors remained in the Press till 1973 to impart training to the Press workmen at different levels. During 1972-73 many operators and apprentices received opportunity to learn the required skill. Moreover, seven Indians were also trained in G.D.R. for 18 months. 17 experienced and trained operators were also appointed for the Process, Photo and Printing Sections who were selected on all-India basis.

The different units of the Press being sophisticated, they need about 200 staff in all including skilled, semi-skilled and unskilled workmen.

To provide opportunity to the local people, special apprenticeship scheme has been enforced so that suitable youngmen of the State with minimum pre-science education having apptitude for manual and mechanical work may carve out their career by joining this course.

Source of Supply of Materials

All the machines in the Corporation have been purchased under quota system at a subsidised rate by the Education Department, Government of Bihar. The Press has not been authorised to purchase any machine without the approval of the Education Department. Other metals and chemicals, however, are purchased from the local market and in case of non-availability of the materials in the local market, they are bought in Calcutta and Bombay. Films for Camera Section are normally purchased in foreign countries on tender basis. Types are casted in the Press through its own super mono-casting machines.

The BSTBPC requires huge quantity of paper for printing of books in different scripts. Therefore, the paper is purchased by the Press through Government agencies under quota system at subsidised rates. However, the Managing Director of the Corporation has been empowered to make purchases up to Rs. 2,000 only. Beyond this limit, the approval of the Education Department and the Board of Directors of the Corporation is essential. About 95 per cent of the total purchase of paper is done under the quota system and the rest 5 per cent from the local market, and that too in the period of exigency.

The Press mainly undertakes the work of printing and publication of text books of Government of Bihar for schools. The number of text books published during 1972-73 was 38 lakhs which rose to three crores forty lakhs during 1981-84. The books published by the Press are of 252 types in 9 different scripts, *i.e.*, Hindi, English, Oriya, Bengali, Maithli, Urdu, Santhali and other regional languages spoken in Ranchi area.

The Press occasionally undertakes the job works of the Government of Bihar and other local offices also.

Trade Policy

Our assessment shows that the Press has no shortage of work. It has so much printing work that about 85 per cent of the work has to be given to outside printing presses, especially to presses in Patna. On inquiry about the allocation of jobs to the private presses of Patna, we came to know that only a negligible number of private presses of Patna receive work from the Corporation and that too not on a regular basis. In 1974, the private press owners demonstrated before the Chief Minister of Bihar and one of the demands was maximum allocation of Government printing work to them. On assurance of job by the Government of Bihar and the then Managing Director of the Corporation, more than 40 private presses of Patna purchased costly Offset machines on loan at higher rates of interest, (*i.e.*, 18 %) from different financial institutions. But the Corporation adopted pick and choose policy in its allocation of work to them. And thus out of the 85 per cent work done outside the Text Book Press, the private presses of Patna have not been able to get even 10 per cent. Therefore, this demands a thorough investigation into the system of work distribution to private presses by the Corporation. Yet one must admit the fact that the BSTBPC owns a strong printing unit supported by a capable set of production equipment and a capable group of working staff. This means that the unit can do better so far as its performance is concerned. Indicators of this good performance may be to judge the quantum of output, its quality, its ability to satisfy the book suppliers as well as other printing presses on a reasonable basis, its effort to minimise the cost of production and its role in promoting harmonious industrial relations as an important public enterprise.

Labour Condition

Staff

The Press is run with the help of the following staff:

Workshop

(*i*) Compositor		
Hand	5	
Machines	3	
		8
(*ii*) Machinemen		12
(*iii*) Fly boys		34
(*iv*) Helper		55
(*v*) Binders		58
(*vi*) Shramik		30
(*vii*) Others		16
(*a*) Technical Officers	4	
(*b*) Assistant Technical Officer	1	
(*c*) Operator including junior operator	9	
(*d*) Works Manager	1	
(*e*) Personnel Manager	1	
Total		213

Office

(*i*) Managing Director	1
(*ii*) Officers, F.A., P.O., A.O., S.O. etc.	14
(*iii*) Accounts Assistants	10
(*iv*) Press Assistants	2
(*v*) Typists	7
(*vi*) Security Guards (including Selection Grade)	18
(*vii*) Peons and Darwans	14
(*viii*) Godown	12
(*ix*) Others	16
Total	94
Grand Total	307

Source: *Unpublished material* (Files), Personnel Officer, Bihar State Text Book Pub. Corporation, Patna-1.

The figures show that the staff composition of the BSTBPC Press is 69 per cent from workshop and 31 per cent from office.

Working Condition

Health and Safety : The BSTBPC has made adequate arrangements for health, safety, cleanliness and welfare facilities. But the Press has no dispensary of its own. All the Press employees get Employees' State Insurance benefit. For this facility they have to contribute a sum of Rs. 15 per month to the E.S.I. Scheme and in return they receive Rs. 18 per day during the period of their sickness in addition to lump sum payments.

Welfare Facilities : The working hours of the workshop staff start from 7 A.M. to 3 P.M. in the first shift and the second shift is from 3 P.M. to 11 P.M. with half an hour recess in both the shifts. The camera section of the Press also functions in two shifts, *i.e.*, first shift from 9 A.M. to 5 P.M. and second shift from 10 A.M. to 6 P.M. On Saturdays the working hours of the workshop are from 7 A.M. to 7 P.M. in the day and from 7 P.M. to 7 A.M. in the night. The office staff of the Press attend their duties from 10 A.M. to 5 P.M. with half an hour recess.

Service Condition

Nature of Employment: The categories of staff employed by the Press are: (*a*) Permanent, (*b*) Temporary, and (*c*) Daily Wage. The total number of employees in these categories is 307 (213 workshop staff and 94 office staff), 50 and 25, respectively. As per our discussion with the Personnel Officer, the Corporation does not put the names of the temporary workers and daily wage workers on Press records. Workers on temporary basis are engaged in the godown of the Press and daily workers are needed during the peak seasons, *i.e.*, March and December of every year.

Main Conditions: The leave rules in the BSTBPC are the same as applicable to the employees of Bihar Government. In other words, Press staff also get earned, casual, medical and extraordinary leave. Medical leave is granted in accordance with the provisions of Employees' State Insurance Act, 1948. Extraordinary leave is granted on the recommendations of the Medical

Officers of E.S.I. Scheme to the Press workers suffering from diseases like lead poison, T.B., etc. The Press staff getting salary below Rs. 1000 avail a lump sum payment of Rs. 500 for their treatment. In addition to above, they also get Rs. 18 per day till they suffer from illness. The Press employees have to contribute a sum of Rs. 15 per month to the Employees' State Insurance Scheme on regular basis for availing the above facilities.

House allowance is granted in accordance with the Bihar State Government rules, *i.e.*, 15 per cent of the basic salary. The Press does not provide quarter facilities to any of its staff. But the employees of the Press have been granted the facilities of loan for house construction. A Press employee can take 100 months salary in advance as loan for this purpose. But the amount will not exceed the limit of Rs. 1,10,000. More than 70 per cent of the Press employees have enjoyed this facility and they have constructed their own houses in Patna.

In case of penalties for offences, warning is the usual system. But strict disciplinary actions are also applied by the Press authorities against the employees committing serious offences. Thus, the Press workers have to face penalties like demotion, bad record, withholding of increment, censure and suspension and dismissals according to the circumstances and the nature of offence committed by them.

Overtime: The Press works overtime regularly. The payment for overtime is linked with the basic salary of employees, at the rate of $3\frac{1}{2}$ hours of overtime equal to a full days duty. However, the Press adopts the following system of calculation for payment for overtime work to its workers:

$$\frac{\text{Basic Salary} + \text{Dearness Allowance}}{162\frac{1}{2}\text{ Hours}} = \text{per hour} \times 2$$

Basis of Promotion: The basis of promotion granted by the Press are test, seniority and efficiency. However, seniority is usually given a higher weightage for promotion purposes. Test and efficiency of workers are taken into account for promotion as secondary factors. In some cases a Press worker is granted promotion on account of having special skill and efficiency irrespective of his length of service.

Grievance Settlement Machinery: Grievances of the employees are channelised through three Unions, *viz.*, Bihar Text Book Kamgar Union, Bihar Text Book Employees Association and Bihar Text Book Karamchari Sangh. The Bihar Text Book Employees Association is a duly recognised Union. Out of all the grievance settlement machineries, Works Committee has been utilised for the settlement of grievances of the workers. The Committee does not function properly these days because of its failure in settling disputes between the workers and the management. And in such a position, the Press workers take help from the Labour Court. Other machineries such as Grievance Committee and Industrial Committee have never been practised for the settlement of grievances.

Facilities: The workers of the Text Book Press enjoy the benefit of compulsory contributory provident fund, gratuity, workmen compensation, bonus and Employees State Insurance. The quantum of benefits regarding above facilities relate to the concerned statutory provisions in the Acts. However, the amount of bonus granted to the employees vary from year to year depending upon the profitability of the Corporation. The workers usually get more than 8.33 per cent as bonus.

Incentive Plans: Sometimes promotion is also treated as a method of incentive plan. Such incentive plans include salary increment, ad hoc payment in cash and rewards in cash and kind. The employees have been awarded with prizes also for their exemplary and efficient performances.

Trade Unionism

There are three labour organisations in the BSTBPC. The Press workers are also the members of these Unions. There is, however, no separate Union of Press workers. These Unions are:

(*a*) Bihar Text Book Kamgar Union
(*b*) Bihar Text Book Employees Association
(*c*) Bihar Text Book Karamchari Sangh

Bihar Text Book Kamgar Union: The Bihar Text Book Kamgar Union was established in 1968 when the Corporation had not started its own press. It was a Union of employees of the publication department of the Corporation till 1971. At present out of the 30 members of the Union, 12 are Press workers. The Union is registered under the Indian Trade Union Act, 1926 and is affiliated with the Centre of Indian Trade Unions. The Sangh receives political support from the Communist Party of India (Marx.). In 1979, the Union had been derecognised by the Press management and since then it has not been properly functioning. The main object of the Union is to ensure adequate wages, secure better conditions of work and obtain fair treatment from the authorities. Negotiations and collective bargaining are two major functions of this Union. On failure of these two methods, the Union favours strikes only. The Union is financed with the membership fee of 50 Paise per month. The leadership of the Union has always been in the hands of the employees of the Corporation.

The Bihar Text Book Employees Association: The Bihar Text Book Employees Association was formed in 1976 for the purpose of protecting the economic, financial and political interests of the employees including Press workers of the Corporation. This association is the result of the failure of Kamgar Union, the oldest Union of the Press employees. The association is duly registered under the Indian Trade Unions Act and its registration No. is 1986/76. It has a total membership of 123 which includes 102 Press staff. This association of Press workers is linked with the Indian National Trade Union Congress and receives political support from the Congress (I) party. The major source of finance of the Union is the membership fee which is Rs. 6 per year with an enrolment fee of Rs. 2 per head. The associations' president and other executives are elected from among its own members. This Union has been duly recognised by the Corporation on the basis of its strength of membership.

The major functions of the association are:

(*i*) to promote cooperation among the workers;
(*ii*) to maintain peaceful relations between labour and management;
(*iii*) to protect the interest of the employees;

(*iv*) to help management in smooth working of the Press; and

(*v*) to organise demonstrations and strikes.

Bihar Text Book Karamchari Sangh: The Bihar Text Book Karamchari Sangh is another association for the employees which includes the Press workers also. The Karamchari Sangh was registered under the Indian Trade Unions Act in the year 1977. Its registration No. is 2026/77. The Sangh is affiliated with All India Trade Union Congress. This association receives support from the Communist Party of India. The activities of the Sangh are to protect the interest of the members and to organise demonstrations. The total membership of the Sangh is 107, out of which 88 members are Press workers. The only source of finance of this Union is the membership fee which is Re. 1 per month. The Sangh occasionally collects some extra amount at the time of strikes. The Sangh has internal leaderships.

The major demands of the Sangh are: (*a*) Revision of pay scale of the helpers, promotion to all categories of Press workers after three years of their service and creation of cadre posts for them; (*b*) payment of 50 per cent of deferred dearness allowance; (*c*) appointment of fully skilled experienced and efficient works managers, electrical engineers and technical officers in the Press; (*d*) revision of scale of pay of experienced, efficient and senior Press staff; (*e*) representation of leaders of the Sangh in the Board of Directors of the Corporation; (*f*) expansion of the Press unit as per advices of BITCO; (*g*) establishment of a branch of a scheduled bank in the premises of the Corporation; (*h*) proper treatment to the Press staff at par with the office staff; (*i*) withdrawal of the Press staff from the office and godown sections; (*j*) establishment of C.P.F. unit in the campus of the Corporation; (*k*) changes in the old standing orders; (*l*) recognition of the Sangh by the management, etc.

Thus, we find that though there are three parallel labour associations in the Corporation, only one association is a recognised body, *viz.*, Bihar Text Book Employees Association. This has the largest number of membership, *i.e.*, 123 out of 307. But it has never played a decisive role of a Union for its members. Our discussion with some employees of the Corporation reveals that the leaders of this Union have to work under pressure from the authorities and they often function as 'yes men'. Since the association does not maintain any records of its charter of demands, we could not evaluate its performance as a representative and effective labour body.

The other associations like Text Book Kamgar Union is still in existence but not actively participating in Union activities. The Union has the lowest number of membership. It was derecognised by the management in 1979.

The third Union is the Bihar Text Book Karamchari Sangh which is the second largest Union of the workers with a total membership of 107. Though the Sangh has not been fully recognised by the management, it is most effective in protecting the interests of the Press workers. It has made a significant contribution to the growth of industrial relations in the Corporation. The Sangh regularly maintains the record of its activities and has got success on several occasions.

Thus, the Press workers of BSTBPC are active members of either of the two associations and they effectively participate in the Union activities.

This much about the Bihar State Text Book Corporation. With this we conclude our survey of public sector presses in Patna. Let us now examine the working conditions in printing industry in the following chapter.

REFERENCES

1. Sharma, M.P.N., 'India's Biggest Printing Press', in *Inauguration of the Text Book Press*, The Bihar State Text Book Publishing Corporation Ltd., Patna, (November 18, 1972), p. 4.
2. Articles of Association, 79(a), Bihar State Text Book Publishing Corporation Ltd., Patna.
3. *Ibid.*, p. 95.
4. Government of Bihar, *Draft Annual Plan, 1984-85*, p. 515.

8

Working Conditions

A survey of working conditions obtaining in printing presses in Patna shall be made in this chapter.

"The conditions under which a person works have got a marked influence on his health, efficiency, psychology and the quality of work he does. It is said that environments create a man, and if we improve the environments we improve the man".[1] It is not possible to carry on the hard work under unhealthy surroundings. An inspiring atmosphere leads to more work than dirty, sullen and unhealthy atmosphere. The attention of the worker is diverted if the surroundings are noisy or dirty. Concentration in the work is most essential and concentration is possible only if there are no outside disturbances to divert the mind of the worker.

In the absence of desirable working conditions, the worker is not comfortable. He then feels as if his job is very tough one. He becomes sluggish and it becomes difficult for him even to pass the time. Good working conditions can remove instability of labour force and can also reduce absenteeism and labour turn-over. The relation between the employers and workers will also improve if the employers take care of the conditions under which the workers have to work.

Working conditions, therefore, govern the efficiency to work. Such conditions largely control the quality and quantity of operations, as also the safety and welfare of the workers. It is for this reason that statutory provisions have been made to follow some minimum standards of working conditions in factories.

Accordingly, an investigation was conducted by us to find out the actual conditions under which the printing industry in Patna functions. Let us now analyse the data collected and information obtained in this field of study under the following heads :

(*a*) Health and Safety,
(*b*) Cleanliness,
(*c*) Welfare,
(*d*) Leave and Holidays,
(*e*) Housing, and
(*f*) Other Facilities.

We shall now take up each of these aspects of working conditions separately.

Health and Safety

A number of factors affect the conditions of health and safety in printing industry. These factors basically govern the capacity and inclination to work and ultimately regulate the health of the workers. They also affect the day-to-day safety of the entire press including its men, machines and materials.

Table 8.1 gives the measures adopted for health and safety in printing industry in Patna.

TABLE 8.1 : **Measures of Health and Safety**

Provisions	Satisfactory		Unsatisfactory	
	No. of presses	Percentage	No. of presses	Percentage
(1) Lighting	91	99.0	1	1.0
(2) Ventilation	90	98.0	2	2.0
(3) Heating	0	0.0	92	100.0
(4) Cooling	0	0.0	92	100.0
(5) Fire prevention	18	29.5	74	70.5

As will appear from Table 8.1 above, the presses under investigation have been classified into two major categories—satisfactory and unsatisfactory. The judgement regarding a press falling in a particular category is based upon : (*a*) the prescribed regulations on the standards of health and safety, (*b*) an on-the-spot study of the prevailing conditions, and (*c*) our personal interviews and discussions with the press workers.

Provisions regarding health and safety in printing industry incorporate a number of measures which are discussed as follows.

Lighting

Table 8.1 shows that about 99 per cent of the presses covered under this research contain satisfactory lighting arrangements. Yet it is important to note that one of the presses works in almost complete darkness. It is Chitra Block and Printing Works.

Ventilation

As regards ventilation, the actual position of the printing presses in Patna shows that about 98 per cent of the presses maintain satisfactory standards. But here too, two presses have poor conditions of ventilation. These presses are Sahni Press and Litho Art Press.

Heating

The position regarding heating is quite depressing in printing presses in Patna. Such heating is required during winter and also for the damp and excessively cold premises. None of the presses, however, have made any provision for proper heating in their premises.

Cooling

The same is the position with regard to the provisions for cooling in the printing presses in Patna. Such cooling is required particularly during the summer when workers find it difficult to work because of excessive heat. But not even one press has made any arrangement for this facility.

Fire Prevention

Protection against fire is a necessity in premises where a number of workers perform their duties with the help of mechanical and electrical equipment. This, however, needs adequate provision for prevention and suppression of fire like exhaust fans, water storage, sand buckets, carbon dioxide cylinders, etc. A reference to Table 8.1 shows that about 70 per cent of the printing presses in Patna do not have sufficient provisions for protection against fire; only 30 per cent of the presses have made some such provision.

Cleanliness

Cleanliness is another factor of efficiency and smoothness of operations. It is more important in printing industry where easy movement of men and materials is a necessity. The Factories Act, 1948 too provides that places of work should be free from drain, dust, refuse, etc., and adequate cleanliness is maintained in and around the place of work. Table 8.2 presents the position of printing presses in Patna as regards cleanliness.

TABLE 8.2 : **Factors of Cleanliness**

Provision	Satisfactory		Unsatisfactory	
	No. of presses	Percentage	No. of presses	Percentage
(1) Sanitation	87	94.5	5	5.5
(2) Drinking water	91	99.0	1	1.0
(3) Washing facilities	87	94.5	5	5.5
(4) Lavatories	87	94.5	5	5.5

Sanitation

Table 8.2 shows that about 95 per cent of the presses in Patna maintain satisfactory standards of sanitation. About 5 per cent of the presses, however, do not have adequate arrangements for proper sanitation.

Drinking Water

The Factories Act expects effective arrangements for adequate supply of drinking water at conveniently situated points in a place where people work. Table 8.2 shows that generally adequate provisions have been made for drinking water in printing presses. About 99 per cent of the presses of the town have made such arrangements at a satisfactory level. But our survey reveals that one of the presses (namely, Madhav Press) has not paid attention to this essential provision.

Washing Facilities

Presses are expected to provide suitable facilities for washing which may include soap, nail brush, tap, wash basin, etc. Our investigation in this respect shows that about 95 per cent of the presses have made arrangements for such facilities and only about 5 per cent of the presses do not have any such arrangement.

Lavatories

Provision for latrines and urinals to be used by the workers has to take into account the factors of convenience, sufficiency and cleanliness. The present enquiry into the working conditions of printing industry in Patna reveals that about 5 per cent of the presses have not made satisfactory arrangements in this area. Nearly 95 per cent of the presses, however, have such facilities at a satisfactory level.

Thus, we find that the condition of the printing presses in Patna with regard to cleanliness may be considered satisfactory. Only a small percentage of presses have neglected this aspect of working condition. This represents satisfactory attitude of the printers towards their workers in relation to cleanliness.

Welfare

Let us now examine the facilities extended by the printing industry in Patna for the welfare of its workers. Such facilities help the workers in the improvement of their living conditions, give the management efficiency of operations and maintain peaceful relationship between labour and management.

Table 8.3 represents the position of the printing industry in Patna regarding labour welfare facilities.

Data contained in Table 8.3 may now be explained in the following paragraphs.

Dispensaries

An investigation was conducted by us to find out the quantum and quality of medical facility available to the press workers in Patna. It was discovered that the workers of printing presses

TABLE 8.3 : **Welfare Facilities**

Provisions	Satisfactory		Unsatisfactory	
	No. of presses	Percentage	No. of presses	Percentage
(1) Dispensaries				
(*i*) First-Aid	41	44.5	51	55.5
(*ii*) Medical Doctors	2	2.1	90	97.9
(*iii*) E.S.I. Scheme	12	13.1	80	86.9
(2) Hours of work	89	96.7	3	3.3
(3) Recess	53	57.6	39	42.4
(4) Other facilities				
(*i*) Cycle stand	7	7.6	85	92.4
(*ii*) Rest Room	5	5.5	87	94.5
(*iii*) Canteen	4	4.3	88	95.7

receive medical services from three major sources : (*a*) First-aid, (*b*) Periodical check-up and consultations with doctors directly appointed by the presses, and (*c*) Benefits under the E.S.I. Scheme.

First-Aid : Adequate equipment for first-aid is a legal necessity for operational units like printing presses. The aid is the most immediate source of assistance to workers in the case of accidents which are so common in mechanical operations. The position of the printing presses in Patna as regards first-aid facility, however, shows that only about 45 per cent of the presses have satisfactory conditions for this type of service. The rest of the presses have either not made any provision or have unsatisfactory arrangements for the purpose.

Medical Doctors : As regards the services of private doctors for periodical check-up and consultations, the workers of only two presses have been able to receive this facility. These presses are: Pioneer Press and Amar Jyoti Press. The rest of the presses, which constitute about 98 per cent of the total, do not have any such provision of receiving medical attention on any organised or regular basis.

E.S.I. Scheme : The Employees State Insurance Act, 1948 includes a scheme for the benefit of the factory workers. These benefits relate to sickness, maternity, disablement, dependence and medical services. Let us first examine the medical benefits under the E.S.I. Scheme. A reference to Table 8.3 shows that only about 13 per cent of the presses have introduced this Scheme of medical benefits to the workers under the E.S.I. Scheme on a satisfactory level. The others, however, have not made adequate implementation of this scheme of assistance.

Hours of Work

A factory has to be run under the prescribed hours of work. This means that both the labour and the management must know as to when the daily work will begin and when will it end. Again, such hours of work have to conform to the prescribed limits for the different categories of the workers. Table 8.3 shows that most of the units in the printing industry in Patna have proper hours of work. About 97 per cent of the total presses may be placed under this category. But three of the presses (namely, Madhav Press, Dariapur, Miscellaneous Printers, Mithapur and Prabhat Press, Mithapur) do not have any fixed hours of work. They work according to their own choice and adjust with the quantum of the job they receive.

Recess

The Factories Act, 1948 prescribes adequate intervals for rest during work. An adult worker is not expected to work for more than five hours without a recess of at least half an hour. But, as will appear from Table 8.3, not all the presses in Patna offer adequate intervals for rest to their workers. In fact, about 42 per cent of the total presses have unsatisfactory provisions for recess to their workers.

Other Facilities

Apart from these major facilities for the workers of the printing industry in Patna, an inquiry was also made into other subsidiary facilities available to the workers. Such other facilities mainly relate to: (*i*) Cycle shed, (*ii*) Rest room, and (*iii*) Canteen.

We shall discuss them separately in the following paragraphs.

Cycle Shed : A number of press workers use cycles for attending to their duties. For the safety and security of such cycles, however, only about 8 per cent of the presses have made satisfactory provision for cycle shed in their presses. The rest of the 92 per cent of the presses do not have any such facility to offer to their workers.

Rest Room : Rest rooms give an opportunity to the workers to recoup their energy lost in course of continuous operations. But probably all the printing presses in Patna are not in a position to make this arrangement. This is the reason why only about 6 per cent of the presses give rooms to their workers for taking rest during their duty hours. About 94 per cent of the presses, however, have no such provision.

Canteen : Canteens supply food and drinks to the workers and thus offer convenience to workers. They also eliminate wastage of time. But organisation or establishment of such canteens is a bit difficult for the printing industry. This is the reason why only about 5 per cent of the presses offer this facility to their workers. The four presses which offer canteen facility are : Atma Press, Bihar Rajya Shikshak Sahyog Sangh Press, the Searchlight Press and the Sanjiwan Press, Patna. However, 95 per cent of the presses do not offer this facility to their workers. In some cases, like a smaller press, this may not be a worthwhile provision too.

This concludes our discussion on welfare facilities in printing industry in Patna. We now pass on to an examination of the conditions of leave and holidays in this industry.

Leave and Holidays

Leave is a system of personal assistance to the workers during the period of their difficulties. Sometimes a worker may be unable to attend to his duties due to bad health or domestic problems. Similarly, holidays are granted by the management on special occasions like festivals, celebrations or days of national importance.

Table 8.4 (also see Fig. 8.1) incorporates the results of our survey regarding leave and holidays allowed by the printing industry in Patna.

The following studies may be made from Table 8.4.

(*i*) Earned leave is granted for a period ranging from 11 days to 15 days in a year by the maximum number of presses in Patna. This, however, accounts for only about 35 per cent of the presses.

(*ii*) Nearly 45 per cent of the presses in Patna do not grant any earned leave to their workers.

(*iii*) As regards casual leave, about 51 per cent of the presses do not have any such provision. This means that the workers of such presses have to attend to their duties even when they are faced with an emergency situation, and if they remain absent from their duties in such a situation, they do not get their pay for the day.

(*iv*) About 42 per cent of the presses, however, do provide the facility of casual leave to their workers. But the number of days available for casual leave during the year varies from press to press. Nearly 14 per cent of the presses allow casual leave up to 5 days in a year, another 14 per cent between 6 to 10 days in a year and the rest 14 per cent from 11 to 15 days in a year.

(*v*) The position regarding medical leave in the printing industry in Patna is rather disappointing. About 79 per cent of the presses do not offer this type of leave to their workers.

(*vi*) The maximum number of presses which offer medical leave is around 10 per cent and they too do not allow such leave to their workers for more than 5 days in a year.

(*vii*) Sometimes workers need extraordinary leave too, but only one press (namely, Bihar Printing Press, Dariapur) has a provision for granting such leave to its workers. The rest of the presses (which constitute about 99 per cent of the total) do not offer such privilege.

(*viii*) Holidays, however, are declared by a larger percentage of presses in the printing industry in Patna. Only about 4 per cent of the presses do not grant any holidays to their workers.

(*ix*) The maximum number of holidays given by the presses are between 21 to 25 days per year. Nearly 61 per cent of the presses allow this number of holidays to their workers.

TABLE 8.4 : **Leave and Holidays**

No. of days	Earned Leave		Casual Leave		Medical Leave		Extraordinary Leave		Holidays	
	No. of presses	Percentage	No. of presses	Percentage	No. of presses	Percentage	No. of presses	Percentage	No. of presses	Percentage
1 to 5	1	1.0	13	14.1	9	9.8	1	1.0	0	0.0
6 to 10	5	5.5	13	14.1	3	3.3	0	0.0	0	0.0
11 to 15	32	34.9	13	14.1	5	5.4	0	0.0	5	5.4
16 to 20	7	7.6	6	6.6	2	2.2	0	0.0	20	21.8
21 to 25	2	2.2	0	0.0	0	0.0	0	0.0	56	61.0
26 to 30	3	3.3	0	0.0	0	0.0	0	0.0	6	6.5
Above 31	1	1.0	0	0.0	0	0.0	0	0.0	1	1.0
No Leave	41	44.6	47	51.1	73	79.0	91	99.0	4	4.3
Total	92	100.0	92	100.0	92	100.0	92	100.0	92	100.0

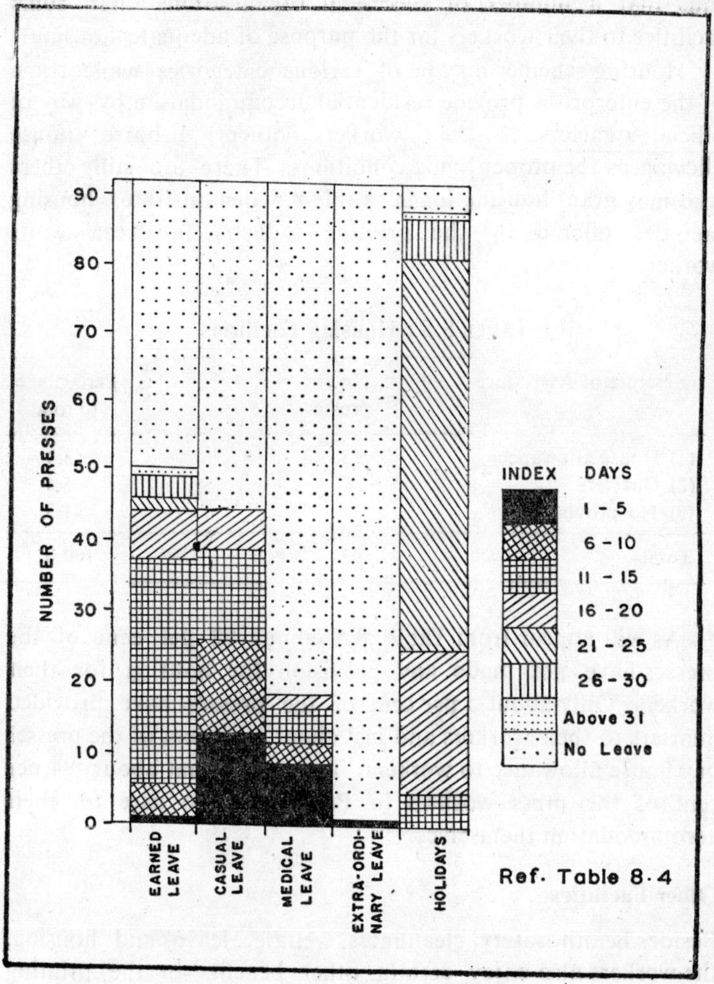

FIG. 8.1 : *Leave and Holidays*

This, in brief, is the position of printing industry with regard to leave and holidays for the workers. Let us now estimate the conditions of housing for the press workers in Patna.

Housing

Housing is a basic necessity of life. It regulates the mental and physical conditions of labour and affects their efficiency. It is thus that a number of successful organisations offer some facilities to their workers for the purpose of adequate housing.

Housing schemes may be of various categories while some of the enterprises provide residential accommodation by way of official quarters to their workers, others disburse house allowances for proper living conditions. There are still others who may grant housing loans. Table 8.5 demonstrates housing facilities offered by the printing industry in Patna to its workers.

TABLE 8.5 : **Housing Facilities**

Nature of Assistance	No. of presses	Percentage to total
(1) House allowance	3	3.3
(2) Quarters	5	5.4
(3) No provision	84	91.3
Total	92	100.0

As will appear from Table 8.5 about 92 per cent of the presses have not made any provision for housing for their workers. Only about 5 per cent of the presses have provided quarters to their workers and just about 3 per cent of the presses pay house allowance to workers. This means that about 94 per cent of the press workers in Patna have to care for their accommodation themselves.

Other Facilities

Besides health, safety, cleanliness, welfare, leave and housing, the workers also enjoy certain other benefits in the printing industry in Patna. Such benefits, in turn, affect the working conditions and initiative to work.

Table 8.6 presents the position of such other facilities in the printing industry in Patna.

The following trends are obtained through an interpretation of the data contained in Table 8.6.

Provident Fund

Provident Fund is a method of financial assistance to the

TABLE 8.6 : **Other Facilities**

Provision	No. of presses	Percentage
1. Provident fund	14	15.2
2. Gratuity	5	5.4
3. Pension	2	2.1
4. Workmen compensation	46	50.0
5. Employees State Insurance	12	13.0
6. Bonus	14	15.2
7. No such facilities	42	45.6

Note : Most of the presses offer more than one facility.

workers. It receives contributions both from the employers and employees for workers benefit. The printing industry in Patna, however, has adopted this scheme of assistance only in about 15 per cent of the units. This means that the workers of about 85 per cent of the presses do not enjoy this facility.

Gratuity

Gratuity is another form of financial assistance granted by employers to employees after retirement. It is paid in lump sum at the time of retirement of a worker. Only about 5 per cent of the presses in Patna, however, offer gratuity to their workers and the rest of the workers of about 95 per cent of the presses do not get this benefit.

Pension

Sometimes workers receive some monthly payment from their employers even after their retirement. This payment is a sort of reward to the employees for their past services rendered to the management. But nearly 95 per cent of the presses do not have any provision for the payment of pension to their workers.

Workers' Compensation

Press accidents are quite common. This results in loss of earning capacity of a worker. The management has to compensate its workers for the financial or physical loss caused by accidents. Only 50 per cent of the total printing industry in Patna, however, follows this scheme of compensation to its workmen. This means that though all the press workers suffer from the same type of risk, not all of them have any uniform system or pattern of compensation.

Employees State Insurance

The Employees State Insurance Act, 1948 provides for a scheme of joint contribution from employers, employees and the Government. This joint contribution is used in granting benefits to the workers for the purpose of sickness, maternity, disablement, dependence and medical relief. This scheme has been implemented by about 13 per cent of the presses. The rest of the 87 per cent of the presses have not adopted the Employees State Insurance Scheme.

Bonus

One of the recent trends in industrial relations has been to encourage workers' participation in management. The workers get a reward for their efficiency in the form of a share in profit considered as bonus. This helps in securing workers co-operation. But only about 15 per cent of the presses in Patna distribute bonus to their employees.

No Such Facility

Table 8.6 further shows that though a number of presses have adopted some scheme or the other for the purpose of rendering financial assistance to their employees, there is a sizeable proportion of the presses which do not offer any such facility to their workers in the form of provident fund, gratuity, pension, workmen's compensation, Employees Insurance Scheme or bonus. About 46 per cent of the presses do not have even a single such facility for their workers. This, therefore, is a slightly depressing trend in the working conditions of the printing industry.

Let us now conclude our investigation into the working conditions of printing industry in Patna.

Conclusion

On the basis of the foregoing analysis, the following conclusions may be drawn in respect of working conditions in printing industry in Patna.

1. Among the measures for the health and safety of the press workers, lighting and ventilation have received adequate attention of the management. But probably

much more is required for the improvement of conditions regarding heating, cooling and fire prevention. Among these neglected aspects of working conditions, fire prevention is a measure of utmost importance.

2. It is encouraging to note that the printing industry in Patna has generally maintained satisfactory standard of cleanliness. On the basis of the study of all the four factors of cleanliness (*viz.*, sanitation, drinking water, washing facility and lavatories) covered under the present investigation, the results obtained show proper care and satisfactory attention in these areas.
3. Our study of the welfare facilities provided by the press management shows that still much has to be done with regard to medical facilities for the press workers. Provisions for cycle shed, rest rooms and canteen may further improve labour welfare.
4. Another important aspect which needs a better attention of the press management is leave and holidays granted to the workers. About 50 per cent of the presses do not grant any earned or casual leave and about 80 per cent do not allow any medical leave to their workers. This means that the press workers have to function under a relatively difficult condition where they are paid only for the job done.
5. A similarly unhappy trend is noticed through our survey of housing facilities for the press workers. Almost 91 per cent of the presses have not made any arrangement or paid any allowance for accommodation to their workers.
6. Most of the presses have failed to appreciate the importance of certain other benefits for their workers. This is particularly true to provident fund, gratuity, pension, Employees' State Insurance and bonus. Just a negligible percentage of the presses provide such other facilities to their workers.

All this, however, brings us to another important area of operation of printing industry in Patna. "A number of labour laws are applicable to the printing industry which create obligations for employers".[2] "These Acts are: Factories Act, 1948; Shops and Establishment Act, 1954; Minimum Wages Act, 1948;

Employees State Insurance Act, 1948; and Employees Provident Fund Act, 1952".[3] The labour laws clearly define obligations of employers as well as privileges of workmen. The translation of employers obligations into workers' privileges is not automatic, but is dependent on the process of implementation of the laws. There are three principal agencies responsible for an effective implementation of statutory provisions. They are: (*a*) Government Department (such as Labour Commissioner), (*b*) Employers, and (*c*) Trade Unions. For Government Departments, these laws prescribe a set of duties and a schedule of work. For Unions, the laws are a source of protection and privilege to their members. The employers treat the laws in terms of direct and indirect costs. For the labour, laws impose certain restrictions on employer's choice and decisions regarding wages, hours of work, condition of work, etc.

In course of our investigation we find that many press owners show their ignorance to statutory labour provisions. Most of the press workers too are not aware of their rights and privileges as granted by the law. This is mainly owing to lack of consciousness or absence of right leadership. This, however, helps the press owners in ignoring the statutory provisions and in denying the required facilities to their workers. The condition becomes still more unfortunate when we find that the Government does not fully recognise printing as an industry. The time, therefore, has come when the management and the Government have to pay full attention to the needs of a fast developing industry like that of printing.

This much about the working conditions. We shall now address ourselves to an enquiry into the labour relations in printing industry in Patna in the chapter which follows.

REFERENCES

1. Saxena, R.C., *Labour Problems and Social Welfare*, K. Nath & Company, Meerut, 1978, p. 604.
2. Thakur, C.P. and Munson, Fred C., *Industrial Relations in Printing Industry: A Study in the Context of Technology and Markets in Delhi Area*; Sri Ram Centre for Industrial Relations, New Delhi, 1969, p. 33.
3. *Ibid.*, p. 22.

9

Labour Relations

THE present chapter is devoted to a study of labour relations in printing industry in Patna.

"Since management essentially consists of getting things done through the efforts of other people, the type of relationship existing between people in an enterprise is the most important single factor in determining how effective that enterprise is".[1] Labour relation is an attempt to secure effective and willing co-operation from employees and reducing conflict between employers and workers and their representatives, *i.e.*, the Trade Unions. It involves Union negotiations, collective agreements and settlement of disputes through conciliation and arbitration. According to John T. Dunlop, "an industrial relations system at any one time in its development is regarded as comprised of certain actors, certain contexts, an ideology which binds the industrial relations systems together, and body of rules created to govern the actors at the work place and work community".[2] In other words, it is the emergence of a pattern of relationship, based not on compulsion, but on voluntary efforts of the parties associated in the enterprise to come closer together, to understand each other, to share the responsibilities as also the fruits of joint endeavour, and to recognise, accept and work towards common objectives.[3] Labour relations emerge largely from recognition of the basic fact that in a free and democratic society the dignity of the individual and moral values should be given the highest place; and authority should be derived from the effective co-operation and active participation of people in the enterprise.

Thus, labour-management relationships are important factors of industrialisation. Promotion, operation or expansion of a trade largely depends upon the attitude of the management towards labour and *vice-versa.* Printing industry, however, has still greater dependence upon labour both for the quantity as well as the quality of the job and its general attitude to work.

With this theoretical background, let us now analyse the findings of our survey on staff employment, industrial disputes, payment, promotion and punishment in printing industry in Patna.

Staff Pattern

Printing industry in Patna has two types of staff, *viz.*, office staff and workshop staff.

Office Staff

Table 9.1 incorporates the types of office staff employed in printing industry, the number of presses engaging such staff and their percentage distribution.

TABLE 9.1 : **Office Staff**

Staff	No. of presses with staff	Percentage to total	No. of presses with out staff-	Percentage to total
(*a*) Manager	45	49.0	47	51.0
(*b*) Superintendent	2	2.0	90	98.0
(*c*) Secretary	1	1.0	91	99.0
(*d*) Assistant Manager	1	1.0	91	99.0
(*e*) Deputy Secretary	1	1.0	91	99.0
(*f*) Assistant Secretary	0	0.0	92	100.0
(*g*) Accountant	15	16.0	77	84.0
(*h*) Clerks	6	7.0	86	93.0
(*i*) Salesman	1	1.0	91	99.0
(*j*) Peon	16	17.0	76	83.0
(*k*) Others (such as Business Representatives and Market Operators)	12	13.0	80	87.0

Table 9.1 shows that about 49 per cent of the presses in Patna have their own managers. But only around 2 per cent of the presses have appointed a superintendent for keeping control over the daily work. The percentage further goes down to just 1 per cent in respect of Secretary, Assistant Manager, Deputy Superintendent and Salesman. None of the presses, however, have any Assistant Secretary. Accountants work in about 16 per cent of the presses, clerks in around 7 per cent and peons in roughly 17 per cent of the presses. Besides these office personnel, some of the presses also maintain their own business representatives and market operators. About 13 per cent of the presses fall in this category. This, in brief, is the position of office staff.

Workshop Staff

Table 9.2 presents the position regarding workshop staff in printing industry in Patna.

As will appear from Table 9.2, the workshop staff consists of several categories of personnel. Let us, therefore, analyse their position separately.

(*a*) *Compositors*—One of the major trends obtained through Table 9.2 is that about 36 per cent of the presses in Patna work with just one or two compositors. The rest of the presses have more than 3 compositors for their job. About 8 per cent of the presses, however, run without compositors.

(*b*) *Machinemen*—Table 9.2 further shows that about 58 per cent of the presses have only one or two machinemen and about 28 per cent of the presses have between three to ten machinemen for their printing work.

(*c*) *Flyboy*—About 55 per cent of the presses do not have any flyboy for assisting the machine operator.

(*d*) *Binders*—Only 29 per cent of the presses in Patna have their own binders. This means, therefore, that about 71 per cent of the presses get their binding work done outside.

(*e*) *Coolies*—Coolies help the subsidiary jobs in presses. But only about 25 per cent of the presses engage coolies for such assistance.

TABLE 9.2 : **Workshop Staff**

No. of workers	Compositors		Machinemen		Flyboy		Binders		Coolies		Proof Readers		Darwan		Others	
	No. of presses	Percentage	No. of presses	Percentage	No. of presses	Percentage	No. of presses	Percentage	No. of presses	Percentage	No. of presses	Percentage	No. of presses	Percentage	No. of presses	Percentage
1 to 2	33	35.9	53	57.6	38	41.2	24	26.0	21	22.8	31	33.7	5	5.4	8	8.7
3 to 4	25	27.2	20	21.7	3	3.2	1	1.0	2	2.1	1	1.0	0	0.0	1	1.0
5 to 6	13	14.1	4	4.3	1	1.0	1	1.0	0	0.0	0	0.0	0	0.0	2	2.0
7 to 8	7	7.6	1	1.0	0	0.0	0	0.0	0	0.0	0	0.0	0	0.0	0	0.0
9 to 10	1	1.0	1	1.0	0	0.0	0	0.0	0	0.0	0	0.0	0	0.0	0	0.0
11 to 12	2	2.2	0	0.0	0	0.0	0	0.0	1	1.0	0	0.0	0	0.0	0	0.0
13 to 14	2	2.2	0	0.0	0	0.0	0	0.0	0	0.0	0	0.0	0	0.0	0	0.0
Above 15	2	2.2	0	0.0	0	0.0	1	1.0	0	0.0	0	0.0	0	0.0	0	0.0
Nil	7	7.6	13	14.4	50	54.6	65	71.0	68	74.1	60	65.3	87	94.6	81	88.3
Total	92	100.0	92	100.0	92	100.0	92	100.0	92	100.0	92	100.0	92	100.0	92	100.0

(*f*) *Proof Readers*—Proof Readers help correction and accuracy of the printed materials. But only about 34 per cent of the presses have one or two proof readers for their business. One per cent of the presses have three to four proof readers. About 65 per cent of the presses, therefore, get their proofs corrected either by the managers or customers.

(*g*) *Darwan*—Sometimes darwans are appointed for the purpose of safety and security of the presses. But just about 5 per cent of the presses maintain either one or two darwans. The rest of the presses have not made any such provision.

Nature of Employment

As regards the nature of employment of the press staff in Patna, the position is as follows:

TABLE 9.3 : **Nature of Employment**

Nature	No. of workers in 92 presses	Percentage to total
(*a*) Permanent	913	78.4
(*b*) Temporary	45	3.8
(*c*) Contract	207	17.8
Total	1,165	100.0

Press workers may be classified on the basis of the nature of their employment in three categories—(*a*) Permanent, (*b*) Temporary, and (*c*) Contract.

Table 9.3 shows that out of a total of about 1165 workers in the 92 presses covered under this survey, about 78 per cent have been appointed to work on a relatively regular basis as permanent, around 4 per cent of the workers are employed as temporary staff and about 8 per cent have been engaged on contract basis. Thus, a major percentage of the press workers in Patna enjoy security of job.

This much about the staff position. We shall now pass on to trade unionism in printing industry in Patna.

Trade Unionism

Printing industry is represented both by the employers' and the employees' associations. Mudrak Parishad is the only association for the printers in Patna. It, therefore, works like a management union. On the other hand, there are two labour organisations for printing industry in Patna. They are—Patna Nagar Press Karamchari Sangh and Patna Press Kamgar Union. The objects and functions of these three associations may be studied separately in the following paragraphs.

Mudrak Parishad

Mudrak Parishad was formed in 1956 for the purpose of guiding the development of printing industry in Bihar. The main office of the Parishad is in Patna and the association has been registered under the Trade Unions Act and the Societies Registration Act of Bihar. The Parishad is affiliated to All India Master's Printers Association, New Delhi.

The major functions of Mudrak Parishad are :

(*i*) to develop a spirit of co-operation, goodwill and understanding among printers in all matters of mutual interest;

(*ii*) to help the printing industry in rendering better and more useful service to the general public;

(*iii*) to maintain peaceful relations between labour and management;

(*iv*) to promote and safeguard the interest of the printing industry and to improve the conditions in all its branches;

(*v*) to develop better methods of sales management and plant management;

(*vi*) to promote sound accounting systems, uniform costing methods, proper ratio analysis and good production techniques;

(*vii*) to promote stability throughout the industry generally by adopting sound business methods, fair wages and salaries and fair trade practices;

(*viii*) to create or manage any trusts or endowments of printing industry;

(*ix*) to establish standard terms and conditions of business in the printing and allied trades;

(*x*) to hold meetings, conferences, conventions and exhibitions;

(*xi*) to publish magazines, periodicals, leaflets and such other literature related to printing; and

(*xii*) to affiliate the Parishad with any other trade or other similar organisations in India and abroad.

Mudrak Parishad, therefore, promotes co-operation among the printers, maintains peaceful relations between the workers and management, and helps management development in printing industry in Patna. It guides adequacy of wages and trade practices and protects them against adverse policies. Thus Mudrak Parishad serves as an important link between labour, management and the Government.

Though the membership of the Mudrak Parishad is open to all printers in Bihar, the actual number of members is just 187. This, however, includes 43 members from outside Patna. This means that out of a total number of 414 presses in Patna, only 144 presses are members of Mudrak Parishad. Hence, some extra efforts are required to expand the membership of the Parishad if it is expected to function as an effective forum of printing industry.

Mudrak Parishad has two types of members. The first category of membership of the Parishad is ordinary members. Any printing press doing business in letter press, litho printing, the die stamping and/or other similar processes of printing, etc. in the State of Bihar is eligible for membership of the Parishad. The membership will be represented by Proprietor, Managing Director, Manager or any representative duly authorised by the press. The membership fee for ordinary members is Rs. 15 annually plus Rs. 5 as admission or entry fee. The second category of membership of Mudrak Parishad is life members. Any press desirous of becoming a life member shall have to pay a subscription of Rs. 501 in one instalment alongwith the usual admission fee of Rs. 5. The Parishad has its patron who is nominated by the Executive Committee of the Parishad on the basis of his overall contribution in the field of printing.

The selection of Patron is made from the list of members of the Parishad.

In course of its operations, however, Mudrak Parishad has made the following major contributions to the development of printing industry in Patna :

(*i*) organisation of two printing exhibitions in Patna during 1959-60 and 1969-70 ;

(*ii*) formation of an Institute of Printing Technology in Patna under its own management for apprenticeship training in printing technology ;

(*iii*) allocation of larger printing jobs to its members from organisations like the Text Book Corporation, Food Corporation, Improvement Trust, School Examination Board, Posts and Telegraphs, etc. ;

(*iv*) participation in the International Printing and Allied Machinery Exhibition, Seminar, Conferences and Conventions held in and outside India (*e.g.*, DRUPA, West Germany, in 1977, IPEX, Birmingham, in 1984 and PAMEX, Delhi, in 1983);

(*v*) co-operation to printers in purchasing offset machines on hire purchase system from N.S.S.I.C. and District Industries Office (Patna) under intensive campaign scheme.

Patna Nagar Press Karamchari Sangh

Patna Nagar Press Karamchari Sangh was established in 1962. The Sangh is registered under the Indian Trade Union Act, 1926 and is affiliated with the centre of Indian Trade Union. The Sangh receives support from the Communist Party of India (Marx.). The main purpose of this association is to attend to the labour problems of the printing industry. Collective bargaining and settlement of disputes are the two major functions of this association. But out of a total number of about 1200 press workers in Patna, the Sangh has a membership of 350 only. The Sangh is financed by membership fee, voluntary donations and occasional collections.

Patna Press Kamgar Union

The Patna Press Kamgar Union is another association for the press workers in Patna. The association was registered under the Indian Trade Unions Act in the year 1970. The Sangh is affiliated with the All India Trade Union Congress. The activities of the Sangh relate to protection of interest of the press workers and organisation of demonstrations where necessary. The total membership of the organisation is 200.

Thus, we find that though there are two parallel labour associations in the printing industry in Patna, there is just one employers' association. None of these associations, however, command sufficient representation of membership. This is one of the basic reasons why none of these associations have been able to make any significant contribution to the growth of industrial relations either from the point of view of labour or management. On personal enquiry from the leaders of the labour and employers' associations we find that they have no strong organisation to represent their interests. The labourers do not actively participate in trade unionism because of fear of termination of their services by the employers. Besides, there is a bureaucratic rigidity of the press proprietors towards their workers. The same situation is faced by the Patna Printers Association. Sometimes printers demonstrate before the Government for the fulfilment of their demands like allocation of more printing works, revision of rate of printing jobs, granting of loans for purchase of modern machines, distribution of paper through quota system, declaration of printing press as a printing industry, etc. The Government, however, has not shown any initiative in tackling their problems in a scientific manner. Arbitrary discrimination in allocation of printing jobs to local presses and lack of understanding and co-operation on the part of the Government have caused enough frustration among the members of Mudrak Parishad. This has further reduced their strength of organisation.

This much about trade unionism in printing industry in Patna. We shall now discuss the methods of settlement of disputes in this industry.

Settlement of Disputes

Table 9.4 shows the methods adopted for settlement of disputes in printing industry in Patna.

TABLE 9.4 : **Settlement of Disputes**

Methods	No. of presses	Percentage
(*a*) Mutual agreement	53	46.5
(*b*) Union intervention	9	7.9
(*c*) Adjudication	13	11.4
(*d*) No machinery	39	34.2
Total	114	100.0

Note : Some of the presses adopt more than one method of settling disputes.

Table 9.4 shows the relative application of various methods of settlement of disputes between management and workers in printing industry in Patna. About 46 per cent of the presses settle their disputes by mutual agreement, about 8 per cent through the intervention of labour Unions and nearly 11 per cent by adjudication. This means that about 34 per cent of the presses do not have any machinery for the settlement of their disputes.

Labour relations in printing industry are also regulated by agreements and relationships existing between the Mudrak Parishad and the two employees' associations. The subjects of negotiations between unions and employers include appointment, wages, hours of work, holidays, overtime and working conditions. The other methods of settling disputes like arbitration, conciliation, mediation, works committee, joint management council, etc., are not prevalent in printing industry.

Payment

To examine the basis of payment adopted in the presses in Patna, Table 9.5 may be presented.

Table 9.5 shows that payments are made by the printers either monthly, weekly, daily or on contract basis. On the basis of the data contained in Table 9.5, we find that a major percentage of presses in Patna make payments to their staff on monthly basis for printing, composing, binding and proof reading. Weekly payments are almost negligible. Only a few presses have adopted the system of daily payments. Contract job is mainly popular for composing or binding work. This means that the printing industry in Patna lacks uniformity

TABLE 9.5 : **Basis of Payment**

Basis	Printing		Composing		Binding		Proof Reading	
	No. of presses	Percentage to total	No. of presses	Percentage to total	No. of presses	Percentage to total	No. of presses	Percentage to total
Monthly	66	71.7	53	57.6	22	24.0	23	25.0
Weekly	1	1.1	2	2.1	0	0.0	0	0.0
Daily	5	5.4	7	7.6	4	4.3	3	3.3
Contract	12	13.1	30	32.7	13	14.1	3	3.3
Nil	8	8.7	0	0.0	53	57.6	63	68.4
Total	92	100.0	92	100.0	92	100.0	92	100.0

of practice in payments. This further holds true when we find from our investigation that only three out of 92 presses grant a scale of pay to their staff. Only three presses (namely, The Searchlight Press, Tapan Printing Press and Pioneer Press) offer periodical increments to their staff and the rest of the presses have an almost stable rate of payment. The wage structure in printing industry in Patna varies from press to press. In big presses like the Government Press and newspaper presses, there is a scale of pay and it varies in three dimensions—time, skill and operation. The scale of pay is followed in a few private printing presses also irrespective of the above three dimensions. But this does not neutralise the cost of living. Moreover, wages in a majority of the private presses in Patna are not only stable, but also too low to provide even proper meals to the workers. In some cases this is an unfair and inhuman treatment to the press workers. A list indicating structure of scales of pay and stable wages of different categories of press employees as per the Minimum Wages Act, 1948, Section 4(1), fixed by the Bihar Government, on 23rd December, 1981 is given in Appendix XI.

Overtime

Need

In addition to the fixed daily wage, press workers also receive payment for overtime work. The need for such additional work beyond the daily hours in presses can be examined through Table 9.6.

TABLE 9.6 : **Overtime Work**

Nature	No. of presses	Percentage to total
(*a*) Required	56	60.9
(*b*) Not required	36	39.1
Total	92	100.0

The figures in Table 9.6 show that overtime work is done in 56 presses constituting about 61 per cent of the total presses. The rest of the presses do not run beyond their normal hours mainly because their quantum of work does not necessitate any such extension of shift.

Frequency

Table 9.7 further explains the position of overtime job in printing industry in Patna.

TABLE 9.7 : **Frequency of Overtime**

Hours	Regular		Occasional	
	No. of presses	Percentage	No. of presses	Percentage
(*a*) 1 to 2	1	5.9	10	25.6
(*b*) 3 to 4	14	82.5	29	74.4
(*c*) 5 to 6	2	11.6	0	0.0
Total	17	100.0	39	100.0

Table 9.7 represents the frequency of overtime operation in printing presses. Out of the 56 presses where overtime work is done, 17 do it on a regular basis and 39 only occasionally. About 83 per cent of the presses run overtime on a regular basis for three to four hours. Similarly, in about 74 per cent of the presses which work on overtime occasionally, three to four hours of extra work is done in a day.

Rate

Rates of payment for overtime job also varies from press to press. This fact is demonstrated by Table 9.8.

TABLE 9.8 : **Rate of Payment**

Basis of calculation	No. of presses	Percentage to total
(*a*) At daily rate	9	16.0
(*b*) Double the daily rate	15	26.8
(*c*) Three times the daily rate	0	0.0
(*d*) Other methods	32	57.2
Total	56	100.0

Table 9.8 shows that about 16 per cent of the presses pay for the overtime at the normal daily rates and about 27 per cent of the presses pay it at double the daily rates. None of the presses, however, pay three times the normal daily rate for their overtime work. But a majority of the presses pay

for overtime as a contract assignment in accordance with the quantum of job produced by the worker. They sometimes also pay the full daily wage for eight hours of work if the worker is on overtime for six hours. Thus, payments for overtime in printing industry in Patna are both on the basis of time rate and piece rate.

Incentive Plan

One of the important areas of our investigation has been a study of the incentive plans for staff efficiency in printing industry in Patna. Table 9.9 incorporates the results of the investigation.

TABLE 9.9 : **Incentive Plan**

Forms of incentives	No. of presses	Percentage
(*a*) Extra increments	32	25.8
(*b*) Prizes and rewards	14	11.3
(*c*) Promotion	14	11.3
(*d*) Other incentives	4	3.2
(*e*) No incentive	60	48.4
Total	124	100.0

Note : Some of the presses provide more than one form of incentive for staff efficiency.

Table 9.9 shows the various forms of incentives offered to press workers and their relative applicability. About 26 per cent of the presses offer extra increments to their good workers. About 11 per cent of the presses offer prizes or rewards when they are pleased with a worker. Promotion in the form of rank, post or scale of pay is granted in about 11 per cent of the presses. Besides these forms of incentives, goods like cycle, uniform, etc., are also given to workers. These other incentives, however, are made by about 3 per cent of the presses. Yet it is slightly regrettable tɔ note that a major percentage of presses do not provide any incentive for outstanding performance.

Promotion Policy

One of the major forms of incentive to staff efficiency is

promotion. The different bases of promotion of press workers are shown in Table 9.10.

TABLE 9.10 **Basis of Promotion**

Methods	No. of presses	Percentage to total
(*a*) Test	17	13.7
(*b*) Seniority	17	13.7
(*c*) Efficiency	60	48.4
(*d*) No procedure	30	24.2
Total	124	100.0

Note : Some of the presses adopt more than one method for promotion of their staff.

On the basis of the data contained in Table 9.10, we find that about 48 per cent of the presses grant promotion to their staff on the basis of their efficiency. Occasional test and seniority are taken into account by about 14 per cent of the presses in each category. About 24 per cent of the presses, however, do not follow any recognised system of promotion.

Penalties and Punishment

Staff discipline is a major problem of management. This, however, needs the formation and execution of appropriate policies both for promotion and punishment. Methods of punishment for press workers, therefore, forms another aspect of our study to which we shall now turn our attention.

T BLE 9.11 : **Methods of Punishment**

Types	No. of presses	Percentage to total
(*a*) Fine	9	6.8
(*b*) Warning	68	51.5
(*c*) Wage deduction	9	6.8
(*d*) Dismissal	13	9.9
(*e*) Bad record entries	9	6.8
(*f*) No penalties	24	18.2
Total	132	100.0

Note : Some of the presses punish their staff by more than one method.

Table 9.11 incorporates the methods of punishment used by the press management in Patna.

Table 9.11 shows that about 18 per cent of the presses do not have any defined system of punishment. Probably they do not give any punishment to their workers. Among the rest of the 82 per cent of the presses, the system of issuing warning appears to be a major practice. In each of the three categories of presses about 7 per cent of the presses impose fines, deduct salary or make adverse entries in the service record with a view to improving staff discipline. About 10 per cent of the presses resort to dismissal of staff for misconduct. But it is the method of warning which is adopted by most of the printing presses in Patna (about 56 per cent of the total).

Conclusion

With this we end our survey of labour relations in printing industry in Patna. Let us now conclude our present analysis:

1. Relationships between labour and management largely govern the operational efficiency of an enterprise. This is more true of the printing industry where both the quantity and quality of job depend upon the attitude and skill of the workers. Moreover, printing industry encourages entrepreneurship among workers. This aspect of the industry has a particular bearing on labour-management relations.
2. Our investigation of labour relations in printing industry in Patna shows that very few presses maintain a good set of office staff for managerial efficiency.
3. Composing is the basis of printing. But not all the presses in Patna maintain a reasonable size of composing section. This adversely affects the quality of production and also restricts employment opportunities of press compositors.
4. Similar is the case with machinemen. Most of the presses are run with a few machinemen which ultimately results in a lower production capacity of the printing industry.
5. Employment of subsidiary assistants like Fly boys, Binders, Coolies and Darwan too fails to show any appreciable trend.

6. A major percentage of the presses do not have their own Proof Readers.
7. It is good that only around 22 per cent of the press staff are either temporary or work on contract basis. The rest of them have security of job by having permanent appointments.
8. There is absence of any organised trade unionism on a satisfactory level. The Patna Printers Association (*i.e.*, Mudrak Parishad) has not been much effective. It was formed in 1956, had a long period of inactivity, was revived in the seventies to put pressure on the Government for revising the printing rates and increasing the allotment of government works, but it has not been successful in any of these efforts. The majority of the presses, therefore, have lost faith in Parishad as an effective forum of printers.
9. The press workers are divided into two Unions controlled and guided by political leaders of different parties. Therefore, the purpose of the Unions is not wholly internal to the Union organisation. The membership of either of the associations is rather low. Thus, both labour as well as management associations in printing industry lack effective representation as a professional body.
10. A sizeable proportion of the presses do not have any standing machinery for the settlement of disputes between labour and management. Yet very few presses go to court of law for the settlement of their grievances. This indicates the general trust and faith of the press workers in management.
11. Most of the presses pay their workers on monthly basis; but whether all of them receive their payment in time may be a matter of further investigation.
12. Only a negligible percentage of presses offer a scale of pay to their staff. In other cases, workers get a fixed salary irrespective of their length of service.
13. Additional work on contract basis is another method used by the employers. It is, however, not widely used in printing, binding and proof reading sections. But a substantial number of presses in Patna use this device for composing purposes constituting about 33 per cent of the

total. Many workers work on daily wage rate during the normal daily hours and accept jobs on contract in the night shift. Thus, contract jobs serve as a supplementary source of income to press employees. Moreover, contract assignments generally accelerate the speed of work and also add to the production capacity of the presses.

14. Overtime work is another means by which the shortage of skill is sought to be relieved. It is practised by about 18 per cent of the presses on a regular basis. Payments for such overtime work, however, are based on both time and job. Thus, overtime also serves as a convenient device for additional income to workers.
15. It is really surprising to note that an expanding industry like printing does not have any sound policy of incentives to efficient workers.
16. It is further discouraging to note that about 25 per cent of the presses do not have any defined basis of staff promotion.
17. It is through warning that most of the cases of indiscipline of press workers is attempted to be set right.

This study of labour relations reveals some gaps in printing industry in Patna. Even if uniformity of practice is not possible in such a large and diversified industrial operation, some recognised principles and policies appear to be essential for management efficiency. Problems of employment, trade unionism, disputes, payments, incentives and punishments, therefore, need a better attention of the press management in Patna to improve the state of labour relations.

Let us now devote our attention to some major problems facing printing industry in Patna in the chapter which follows. This will serve as a fore-runner to our final conclusions.

REFERENCES

1. Indian Institute of Personnel Management, *Personnel Management in India*, Asia Publishing House, New Delhi, 1976, p. 83.
2. Thakur, C.P. and Munson, Fred C., *op. cit.*, p. 3.
3. Indian Institute of Personnel Management, *op. cit.*, p. 84.

10

Problems of Development

THE object of this chapter is to identify the major problems facing the development of printing industry in Patna.

To organise and operate an industry efficiently is a managerial challenge. This is more so for printing industry which mainly depends on a balanced relationship between men, money, management and machine. Moreover, printing industry is much different from other industries. The industry consists of printing presses, large in number but small in size. Many old presses collapse but new presses come into existence in bigger number irrespective of their choice or knowledge of printing trade. This abrupt growth in the number of printing presses adds new dimensions to the problems of printing industry.

The demand for and the supply of printing materials have their own peculiarities. While other industries can produce and stock in advance in anticipation of future demand, printing commences only when the required order has been received. "Printing cannot be produced in anticipation of demand: a house may be built as a speculation and may well sell at a considerable profit, but it would be a rash printer who would produce catalogues and posters without knowing the requirements of his customer".[1] It may be easier to print forms and account-books in anticipation of consumption, but it is not possible to produce catalogues, pamphlets, visiting cards, invitation cards, letter heads, or posters to suit the requirements of haphazard customers.[2] Moreover, the process of production and distribution may not be able to receive any substantial assistance from the techniques of business forecasting in printing industry. Here the nature of relationship between the customer and the printer also differs from other industries. For it needs time to print and each print has to meet the specific taste and requirements of every customer. "Printing, therefore, is a complicated process of multiple steps and requires a wide range of related equipment".[3]

"An Indian press owner has to encounter many problems and face tremendous odds in his daily business".[4] The problems are divergent in nature. They affect the plans as well as policies of management and the task of trade management becomes an extremely difficult one. Such problems restrict both the quality and quantity of operations. The problems of printing industry, therefore, need adequate attention for effective regulation and control of the business. Although the problems of printing industry are many, they are not so difficult provided they are handled tactfully.

The problems of printing industry may originate from the following areas:

(*a*) Government policy;
(*b*) Finance limitations;
(*c*) Customer's choice;
(*d*) Labour relations;
(*e*) Market competition.

The problems are further aggravated by: (*i*) Absence of trained manpower, (*ii*) restrictions of foreign exchange, (*iii*) scarcity of paper, (*iv*) increasing cost of printing materials, (*v*) excessive rate of depreciation of types and printing equipment, etc. All these jointly affect the productivity, efficiency and economy of operations in printing industry and its balanced growth.

Let us now examine the nature, variety and dimension of these problems in detail under the following heads:

(*a*) Capacity Utilization,
(*b*) Government Policy,
(*c*) Labour Efficiency,
(*d*) Mutual Co-operation, and
(*e*) Economic Operation.

We shall analyse these problems of development of printing industry separately.

Capacity Utilization

Whether there has been inadequacy of work to feed the printing capacity of this industry in Patna has been an important area of our investigation. It is in this perspective that the problem of capacity utilization receives attention. The problem has two major aspects: whether the goods are delivered in time and meet the quality standards, and whether the printing job in the area is surplus enough to be entrusted to outside presses. When an attempt was made to make a deeper probe into these questions through personal enquiry from the press owners, their representatives and the officials of the Government of Bihar, it was discovered that capacity utilization was a key problem of the industry. This, however, includes a number of such problems which have not received the attention they deserve. Let us now see those problems.

Inadequacy of Work

Our investigation has shown that many presses in Patna remain idle for varying lengths of time. Table 10.1 shows the percentage of presses facing inadequacy of work.

TABLE 10.1 : **Inadequacy of Work**

	No. of presses	Percentage to total
Yes	68	73.9
No	24	26.1
Total	92	100.0

As will appear from Table 10.1, about 74 per cent of the presses suffer from shortage of work. The reasons for this inadequacy have been explained in Table 10.2.

Table 10.2 shows that the following four major causes are responsible for inadequacy of work in printing industry in Patna:

(*a*) Limitation of job;
(*b*) Shortage of funds;
(*c*) Seasonal fluctuations; and
(*d*) Unfair competition.

TABLE 10.2 : **Causes of Inadequacy of Work**

Reasons	No. of presses	Percentage to total
(*a*) Limited work	59	21.5
(*b*) Bad organisation	3	1.1
(*c*) Shortage of capital	57	20.8
(*d*) Lower quality of work	0	0.0
(*e*) Seasonal fluctuation	61	22.3
(*f*) Intense competition	59	21.5
(*g*) Other factors	35	12.8
Total	274	100.0

Note: More than one factor affect the capacity utilization of a press.

The period of crisis for the industry, however, varies from press to press. Table 10.3 demonstrates the degree of inadequacy of work during the different months of the year.

TABLE 10.3 : **Periodical Distribution of Work**

Months	No. of presses	Percentage of total	
(*a*) January	0	0.0	
(*b*) February	1	0.6	
(*c*) March	8	4.6	
(*d*) April	23	12.9	
(*e*) May	25	14.1	
(*f*) June	30	16.9	69.9% (April–August)
(*g*) July	25	14.1	
(*h*) August	21	11.9	
(*i*) September	7	3.9	
(*j*) October	4	2.3	
(*k*) November	1	0.6	
(*l*) December	0	0.0	
(*m*) Always	32	18.1	
Total	177	100.0	

Note: Some of the presses have inadequacy of work during several months of the year.

Table 10.3 incorporates the major trends of work shortage in printing industry in Patna. Though about 18 per cent of the presses complained about constant inadequacy of work, around 70 per cent of the presses face this problem only during the months of April to August. If, therefore, efforts are made to feed the presses during the dull season, it will give a better utilisation of production capacity in printing industry.

Unfair Distribution

There is a feeling in some quarters of the industry that the printing job is not distributed fairly. A few big presses consumed a larger share and the rest either do not get any job or get only an insufficient allotment. This results in over-work for a few and starvation for others. But this feeling of the trade can hold true only in the case of government work. Even then, a government agency is free in its choice. For the government work also, there is a separate government press, and none of the private presses are expected to depend entirely upon government business.

Unhealthy Competition

Very few problems have affected printers as acutely as unhealthy competition between one printing press and another in the same locality or in different regions. This problem has put a very great strain on small and medium size presses. Although competition is a boon to consumers, it is not so for the industry. When presses compete with each other, the charges go down. But it is always at the cost of quality. The problem of problems, therefore, is how to initiate competition in printing industry for a better quality of production rather than for a lower cost so far as the interests of the industry are concerned. Whether a customer needs quality work even at a slightly higher cost may be another area of study which will need further investigation.

Large Number

An important question before the industry is—in view of inadequacy of work where is the need for additional presses? When the existing presses themselves are not able to utilize fully their capacities, why to establish new presses? Does it mean that some restrictions should be imposed upon the new entrants to the industry? And if so, what restrictions and how? If at all such a restriction appears to be essential in the interest of the trade, who will be the appropriate authority for the purpose—Mudrak Parishad, Government, Trade Unions, or any similar other agency? These are some of those questions which need attention of the printing entrepreneurs.

Delayed Payment

One of the main factors contributing to the early winding up of presses is delay in the settlement of their dues by the customers—be they the government or a private party. Timely payments are necessary to keep an industry alive as also to develop it further. It is promptness of payments which generates the required funds and offers liquidity of resources in printing units.

Poor Equipment

A number of presses in the town owe their existence to accidental entrance rather than a planned effort. This has resulted in the absence of proper plants and equipment in most of the presses. Either the presses do not have the right type of machines or the condition of the machines is not satisfactory. Both of them, however, badly affect the quality as well as the quantity of production. This further results in jeopardising consumer's satisfaction and timely delivery of work. Thus, if the industry has to perform its job smoothly, it must be equipped with proper plant and machinery.

Technical Know-how

Not only that, even the technical know-how available to the trade is in a deplorable condition. The press workers, be they compositors, printers, binders or even managers, have come to serve the industry more as a source of livelihood than a line of specialization. This is one of the reasons why a number of representatives of the presses in Patna pleaded for appropriate training in apprenticeship programmes for the workers of this industry.

All these problems hamper the process of effective utilization of printing capacity in Patna. This is mainly due to a mushroom growth of printing presses, on the one hand, and their unplanned organisational set-up, on the other.

Government Policy

Most of the printers whom we met in course of our investigation complained about State indifference towards this industry.

It is said that the government itself has been responsible for a number of problems. The printers feel that a sound industrial policy is the need of the hour to protect the printing sector. The government must come out with its plans to develop printing industry. Let there be a more defined responsibility of the government for the desired industrial development in printing sector. The attitude and approach need modification so as to initiate better co-operation. One issue which now needs examination is—how can the State help the industry?

The issue was discussed in depth by us at a group level meet of the printers of Patna specially invited for the purpose. The meet was represented by each level of printers who examined the different problems of printing industry on the basis of a questionnaire specially designed by us for the purpose (Appendix XII). It was found that the major need of the hour is the setting up of a separate Department of Printing by the Government of Bihar. This department should render alround assistance to this industry for its fair operation. The Department of Printing should have an advisory body consisting of representatives from the press management, government, trade unions, publishers, statutory boards or corporations and experts who have made some study of this industry. The following may be the possible lines of contribution to be made by the proposed Department of Printing.

Distribution

There should be a fair distribution of the printing work. Normally such jobs should not go outside the State. Adequate allocation of printing work should be made between: (*i*) small and big presses, and (*ii*) government and private presses. This type of distribution, however, is needed for the work of government and semi-government agencies.

Quota System

Of late, there has been an alarming scarcity of paper in the country—paper being the basis of printing. The private presses in Patna face severe shortage of paper from time to time. The government has not shown any helping attitude to the private presses in this direction. Majority of the presses get supply of paper from the local market. Thus, the supply of paper to the press depends upon their availability in the market. If there is no paper in the market, the presses in Patna have to sit idle for months. Moreover, there is no set mechanism for the distribution of paper either from the government or the Mudrak Parishad. The different presses, therefore, should have prescribed quota of paper according to their capacity of work and quantum of consumption.

Finance

Adequate financing of industrial development is a necessity. However, our survey has shown that a major percentage of presses are ill-equipped, unorganised and mismanaged. The management is not able to obtain proper machinery mainly due to lack of funds. Presses, therefore, need financial assistance in the form of loans and advances from time to time.

Training

The problem of training to press workers is no less important for industrial efficiency. But there has not been any organised attempt on a satisfactory level for this purpose. If the Department of Printing can sponsor some training institute or some orientation programmes for press workers, they can contribute to both labour efficiency and management profitability.

Rates

The printing rates charged by different presses widely vary. Even for the government job tenders have to be invited time and again. So, why not lay down some standard minimum rates of printing? Obviously, this should be based on the size of the press and the quality of its production.

Regulations

This research has experienced both ignorance as well as indifference of the printers to the regulatory laws governing the printing industry. A printing press is governed by two sets of laws. The first relates to the declaration and registration of presses and the second to labour legislation regulating the welfare and working conditions. For starting a business, a certificate of declaration from the government is a must. Accordingly printers in Patna follow the rules of commencement of business and obtain certificate of declaration, for running a press. Like declaration, the registration of a press is also essential. But it has been found that about 31 per cent of the presses in Patna have not got themselves registered under any of the Acts, *e.g.*, Factories Act and Shops and Establishments Act. This shows the ignorance, indifference and negligence of the presses in Patna to the statutory framework.

So far as the implementation of other labour laws is concerned, the organised and big presses meet the statutory requirements of these Acts and enjoy their benefits. But most of the presses being small in size, are excluded from the operative effects of these Acts and other benevolent enactments. This is owing to statutory limitations imposed in the Acts for lesser number of men or machine employed by printing presses being kept out of the purview of law. This needs a reorganisation of the different rules and regulations framed for the purpose of controlling the operation of both big and small printing presses. There should be a clear-cut direction in respect of declaration and registration of presses, sound working conditions, adequacy of equipment, fair wages, etc. Only this can protect the interests of labour, management, government and the customers.

Research

Printing is an important activity. It provides the basis for expansion of education and information. But not much has been done for the collection of data, analysis of trends, study of problems and attention for growth of this industry. This is mainly because the industry is pre-eminently composed of small firms. Researchers have hitherto neglected the small-scale industries sector. Hence, proper research regarding printing industry may identify the right lines of its further development.

A Department of Printing, therefore, appears to be a necessity. The functions of this Department may be on the above lines. In fact, some other States like Assam, Haryana, Maharashtra and Bengal are already having a department of this nature. This means, therefore, that printing should not remain only as a trade. It should be recognised as an industry for the purpose of granting protection and assisting expansion. This type of government policy is one such major problem which does not need much elaboration.

Labour Efficiency

No enquiry into any industry will be completed without a detailed discussion of the problems of labour. Printing industry is no exception to it. Labour problem have many facets such as labour conditions, labour remuneration, industrial relations and so on. These varied problems of labour directly affect their efficiency to work. The problems of labour efficiency in printíng industry in Patna, therefore, is another aspect to which we shall now address ourselves.

Operational results are governed by the quality of labour. It is all the more important in the printing industry where men and machine have to function as close to each other as is possible. But efficiency of operations is not an absolute achievement. It needs several considerations. Some of the problems which affect efficiency of labour in this industry may be presented in the following manner.

Employment

Presses employ labour both on contract as well as on regular basis. Where contract system is followed, quality is badly affected, although the quantum of output may be slightly higher. Where regular appointments have been offered, labour gets a minimum rate despite the quantum of output. It, therefore, needs sound judgement as to which of these two systems will contribute to labour efficiency.

Training

Only a few press workers possess adequate training. A majority of them have come to this line as a source of livelihood. They are, therefore, entrants by chance. But it is the latter category which dominates the press labour. They have developed experience and they normally complete their work with reasonable speed and quality. Whether training from recognised sources is more conducive to the growth of labour efficiency in printing industry or experience of work alone can provide the required specialisation can be another area of research investigation.

Migration

A number of workers change not only their line of operation but also the organisation. Sometimes they work as compositors, sometimes as machinemen and sometimes as binders. Moreover, they do not stick to one press. They migrate from press to press. Therefore, it is a big question to be answered whether the migratory character of workers help labour efficiency in printing industry.

Equipment

A good number of workers who were interviewed in course of this investigation expressed their difficulties owing to inadequate printing machines and equipment. Maximum efficiency is not possible in the absence of proper materials. It is, therefore, surprising to expect adequate efficiency from workers when the management itself fails to provide them with proper machines in normal working order.

Location

Most of the presses which were visited by us present a gloomy picture in respect of working conditions. Presses are located in dirty lanes. There is no space to move freely. With the exception of some large presses, most of the regulated and unregulated presses, especially the latter, are generally located in premises which were designed for purposes other than those of housing a factory. It is not surprising, therefore, that press workers are working in "stables", improvised sheds, cow sheds and ill-ventilated old houses having dark rooms. In many presses walls and ceilings are seldom white-washed or painted and have accumulated a heap of dirt and cobwebs. These presses are having the problems of over-crowding and congestion. The flooring even when cemented and brick-laid, remains dirty, uneven and unclean. Due to this, discomfort of congestion increases. Besides, high temperature in summer, cold in winter and leaking roofs in the rainy season are the normal features of most of the presses in Patna. The paucity of fans in most of these presses can hardly be considered conducive to the efficiency of the workers. These are combined with concentration of presses in a few areas. Would it, therefore, be fair to talk of labour efficiency when even normal conditions of work are not available to press workers?

Responsibility

Another factor to which attention has to be paid is how to judge the efficiency of labour in printing industry. If it means a good production, then who is more responsible—compositor, printer or manager? The compositor has to show accuracy combined with speed. The printer regulates the impression. But is not the management responsible for the overall efficiency of production? Obviously this has to be represented by a correct and clear impression in the printing job. It is management which checks up the proof and finalises the print order. This means, therefore, that labour efficiency is not one-way traffic.

If these problems of labour efficiency are analysed further, one would find that the problem at base is not efficiency of labour, but of a coordination between labour and management. Unless both work together, the quality or quantity of production in printing is not possible. It is this type of coordination between the two which demands attention in printing industry in Patna.

Mutual Co-operation

If individuals are faced with certain problems which they cannot solve all alone and the government also fails to come to their rescue, perhaps a co-operative effort of all concerned remains to be the only alternative solution. Co-operation is a pre-condition to successful management. It is, therefore, equally important to the printing industry. Industrial operations are faced with a number of problems. If co-operative efforts can be made to solve such problems, it serves the cause of business growth as well. The suggested areas of co-operation in printing industry, however, may be as follows:

(*a*) Co-operation between the printers;
(*b*) Co-operation between the workers;
(*c*) Co-operation between the workers and the management;
(*d*) Co-operation between the workers and the government;
(*e*) Co-operation between the management and government; and

(*f*) Co-operation between the workers, management and the government.

Let us now examine these areas of co-operation in detail.

Printers' Co-operation

To find out exact answers to the problems of printing trade and industry, we held several group level meets of printers, skilled workers, leaders of Mudrak Parishad and the Superintendent of Government Printing Press. Problems of co-operation were discussed in detail. At the meet it was generally agreed by the participants that there should be a printers' co-operative. It was felt that the co-operatives should attend to the common problems of the industry. The possible lines of contribution of the printers' co-operative may be:

(*i*) to assist fair distribution of essential commodities for the industry, such as paper, metal, ink, etc.;
(*ii*) to provide a workshop for the repair of machines and equipment used by the presses;
(*iii*) to help a balanced distribution of work among the various printers;
(*iv*) to introduce labour welfare schemes for personnel development;
(*v*) to standardise the printing rates with uniformity in charges and quality of operation;
(*vi*) to render technical supervision through periodical visits of co-operatives' mechanics;
(*vii*) to organise a store for essential spare parts often required by the industry;
(*viii*) to sponsor training programmes for those who are engaged in this industrial activity; and
(*ix*) to obtain credit facilities for the printers through the recognised sources of financing.

The proposed Printers' Co-operative, therefore, may aim at reduction of unhealthy competition and promotion of mutual co-operation in printing industry in Patna. It may work for the common interest of the trade rendering both advisory as well as executive services. It may be designed to protect the interest of the industry to help its further development and to facilitate its smooth operations.

But whether the existing association of printers, *i.e.*, Mudrak Parishad, may be able to manage all such activities or there is the need for a separate co-operative on the above lines is another problem to which we shall devote our attention at a subsequent stage.

Co-operation between the Workers

It is needless to emphasize the importance of co-operation among the workers of the same industry. They share their feelings and experiences. Sometimes they have common problems. Therefore, if they can work together they may be able to participate more effectively in industrial efficiency. This may form a workers' co-operative in printing industry.

Co-operation between Workers and Management

Both printers' as well as workers' co-operatives have to be close to each other. Co-operation between the printers or between the workers is meaningless until the two present a common participation in the industrial development programmes. When workers co-operate with the management, both labour satisfaction and efficiency go up. It is, therefore, necessary that both labour and management co-operate with each other for solving the industrial problems.

Besides these areas of co-operation, there is also the need for co-operation between: (*a*) the workers and the government, (*b*) government and management, and (*c*) workers, management and the government jointly for an effective effort towards development of printing industry.

Economic Operation

Economic control of an industry is a difficult task. A number of problems restrict the economic progress of an industry. Printing industry is no exception to it. The two major problems in the economy of the printing industry are: (*i*) finance, and (*ii*) printing cost.

Finance

Our investigation into the financial structure of the printing industry in Patna shows that inadequacy of resources has been a major limitation to its progress. Table 10.4 incorporates the results of our investigation in this area.

TABLE 10.4 : **Source of Finance**

Capital	No. of presses	Percentage to total
(*a*) Individual	89	72.9
(*b*) Corporate	3	2.4
(*c*) Bank loan	13	10.6
(*d*) Private money-lender	10	8.2
(*e*) State assistance	6	4.9
(*f*) Debenture/Bonds	1	1.0
(*g*) Institutional support	0	0.0
Total	122	100.0

Note: Some of the presses are having more than one form of capital.

Table 10.4 shows the following trends in the financing of printing industry in Patna:

(*a*) About 73 per cent of the presses rely upon individual sources of capital.

(*b*) The contribution of corporate capital to the financing of this industry in Patna is just around 2 per cent of the presses.

(*c*) As regards borrowing, bank loans contribute the largest share of the industrial requirement.

(*d*) Private money-lenders finance this industry to the extent of about 8 per cent of the total presses.

(*e*) Contribution of State assistance to the financing of printing industry is almost negligible.

(*f*) None of the presses receive any assistance from specialised financial institutions.

These trends show the relative importance of the different sources of finance available to printing industry in Patna. In view of the present inadequacy of resources, the problem is to augment the funds of the industry; funds which may help improvements in the quality and quantity of printing.

Printing Cost

It is said that cost is a fact and price is a policy. Cost study is now a pre-condition to industrial development. Like other industries, cost control and economic operation are also necessary in printing industry. The printer, therefore, should exercise control over costs and, for this purpose, he must have proper information flow about the different elements of costs of production incurred during the process of printing. Moreover, the printers must be in a position to avoid all wasteful expenditure. All staff should examine critically the day-to-day performance and the costs incurred keeping in view the prescribed standards.

Thus, a sound costing system is pre-requisite to successful functioning of printing industry. Even the British Federation of Master Printers has endorsed the need for proper costing system in printing industry. In fact, it serves as a basis to price policy. It is, however, slightly surprising to note that the printing industry has not adopted uptill now any uniform system of costing. Table 10.5 gives the position of costing methods adopted by printing presses in Patna.

TABLE 10.5 : **Methods of Costing**

System	No. of presses	Percentage to total
(*a*) Presses having some standard systems	4	4.4
(*b*) Average Market Rate	14	15.2
(*c*) No system	74	80.4
Total	92	100.0

From Table 10.5 it is obvious that more than 80 per cent of the presses in Patna do not follow any recognised principles of costing in printing. They often charge for the goods and services not on any scientific basis of estimates, but according to their own arbitrary decision. Especially, the new presses come under this category and they charge below the average price ruling in the market.

15 per cent of the presses too do not have any sound method of costing; rather they depend upon the average market rate. Only 5 per cent of the presses can be considered to have adopted some standard norms of pricing of printing services. This absence of a uniform system of costing in printing industry has resulted in a number of problems, such as:

(*a*) Different rates of printing charged by different presses for the same job.
(*b*) Printing charges varying from time to time even in the same press.
(*c*) Competition in quotations for printing jobs.
(*d*) Excessive bargaining by customers for printing charges.
(*e*) Difference between the quality agreed and the quality produced.
(*f*) Disproportionate profit or loss to the press organisation.
(g) Loss of confidence by the customers.
(*h*) Difficulties in preparing correct estimates.
(*i*) Difficulties in variance analysis of standard costs and actual costs.
(*j*) Inadequate financial management in printing industry.

The need for an adequate system of costing in printing industry, therefore, is obvious. In fact, costing is essential to the efficiency and effectiveness of operations. But probably just a few presses have recognised the need for proper costing systems in industrial management.

Our further probe into this problem has shown that either the presses have not realised the importance of costing or they have not been able to develop the correct technique of costing. At this stage it may be worthwhile to design a costing system appropriate to printing industry and suggest its adoption at the earliest.

This shows the different elements of cost of printing. Modification in the cost structure, however, may be necessary to suit organisational requirements. This may facilitate calculation of cost, determination of profit and ascertainment of printing charges.

Estimate of Printing Cost

Elements		*Amount*
1. *Labour*	(*a*) Composing	....
	(*b*) Proof Reading	....
	(*c*) Printing	....
2. *Materials*	(*a*) Printing ink	....
	(*b*) Paper	....
	(*c*) Printing types	....
3. *Management*	(*a*) Depreciation of machines	....
	(*b*) Electricity and power	....
	(*c*) Office Expenses	....
	(*d*) Repairs	....
	(*e*) Interest	....
4. *Additional services*	(*a*) Cutting	....
	(*b*) Binding	....
	Total	....

With this we end our study of major problems of printing industry in Patna. The study has focused attention on capacity utilization, government policy, labour management, mutual co-operation and economic operation. Approaches to solve these problems would be discussed in the concluding part of this study. The present analysis, therefore, serves as a fore-runner to our conclusions.

REFERENCES

1. Montague, Noel, *Printing Management*, Sylvan Press Ltd., 24 Museum St. London W.C.I., 1948, p. 175.
2. *Ibid.*, p. 119.
3. Thakur and Munson, *op. cit.*, p. 62.
4. Saifuddin, T.A.. *op. cit.*, p. 7

11

Conclusion

WE have now reached the concluding part of the present research on printing industry in Patna. Our findings and suggestions may be summarised as follows.

Printing Skill

Printing is one of the most developing skills in modern age. It is a powerful technique of mass communication. It is, therefore, an important aspect of the economy. Printing industry is one of the largest industries in India absorbing millions of skilled and unskilled workers. But unfortunately, it has failed to get the recognition it so highly deserves and its true role in the economic structure of the country is yet to be appreciated. Even a systematic study and evaluation of the problems facing this developing industry has seldom been made by the social scientists. The statistical organisation in respect of information flow for printing industry too is rather weak. This makes it difficult for anybody to know even the number of printing presses in India, their employment structure or their regional distribution. This gigantic industry itself has failed to organise effectively. It has thus failed to receive the required attention from the Government as well the trade. Nevertheless, the efficiency of this industry depends not on the quantum of output (*i.e.*, the number of print impressions) but on the quality of its production. Probably the latter warrants greater attention in a competitive economy. It is this which has made management of printing industry a difficult task, particularly in the absence of the required patronisation from either of the quarters responsible for the development of a highly labour intensive industry like that of printing.

Our investigation into the evolution and growth of printing industry reveals that printing, basically known as reduplication, is as old as the human civilization. It has its different stages of growth. Each stage, however, has its own characteristic features. Whatever types of printing are available today, block form has been the first and the foremost. This type of printing has been traced back to China in 650 A.D. with portraits of Buddha. This was further supplemented by Sir Rurel Stein in 1907 when he discovered "Hiraka Sutra" as the oldest printed book in the world. In Europe block printing was known much earlier than the use of movable types in fifteenth century. In India printing has been known since ancient times. Five thousand years old pictures have been found in the excavation of Mohenjodaro. "Mudra" (Seal) appears in Kautilya's Arthasastra between 100 B.C. to 400 A.D. These facts prove that printing in old form has been used in India since long. However, the beginning of printing differs from country to country and it has spread over about a century from 1455 in Germany to 1593 in Philippines. In India printing was known in 1577.

With gradual expansion in the socio-economic and political requirements, printing industry in India got further support. Consequently in the year 1797 types of different Indian scripts were manufactured for the purpose of translation of New Testament. Thus, the importance of printing was badly felt by Indians. But it could not continue for long and unfortunately our printing industry did not receive its due importance from the Government as well as from its own organisation. The industry, therefore, could not form part of the planning process of the Government. Whether it is literature on the subject or report of the Government or All India Master Printers Report on Printing, stories of neglect are rampant. It is surprising that in spite of its educational, economic and commercial contributions to the process of development, printing remained as a disorganised industry.

Needless to add that the earliest a Central Coordination Committee on Printing is set up in the country, the better it is.

The proposed Committee should be responsible for the identification of the problems of printing industry in India and for the recommendation of suitable measures to improve its status.

Organisational Characteristics

Our enquiry into the state of printing industry in Patna shows that there are in all 414 presses in the town. They represent a growth history of about one hundred and five years. But just a few presses have been able to develop into sound organisational units. Most of the presses, therefore, lack modern equipment, function under poor conditions and face inadequacy of resources.

A major technique of the present survey has been collection of information from primary sources. Accordingly, our findings about the conditions obtaining in the units selected for intensive study cover 92 private presses and 3 public presses. Though the data contains an analytical study of 95 presses, it is a truly representative picture of the entire industry demonstrating clearly its nature and problems.

A study of the development of printing industry in Patna shows that it has missed an organised growth. This means that a planned promotion or expansion of this industry is absent. This has caused erratic establishment as well as abrupt closure of many presses.

Our investigation demonstrates that there has been a gradual expansion in the equipment growth of this industry. The printing industry in Patna has rapidly shifted from hand press (a traditional system) to automation-process of printing. But not a substantial number of presses have followed the process of automation which indicates a discouraging state of affairs of a large number of presses in the town.

The study of job specialisation of printing industry in Patna reveals that the growth in the modern printing equipment has led to an expansion in their job specialisation. It is encouraging to note that the presses in Patna have not confined to one job, rather multi-job systems have been undertaken by them. The process of specialisation in printing industry, however, has seldom followed any planned effort. Neither the government nor the printers' organisation issues any guidelines to presses for a scientific or planned job specialisation. This has often resulted in a haphazard growth of printing operations. It may, therefore, be suggested that the Mudrak Parishad should take additional initiative by guiding the destiny of the presses and by counselling them on job specialisation.

Time Study

In course of our investigation, we conducted a time study of three presses with a view to judging their performance evaluation. Though there is no single criterion which can serve as an effective indicator of press efficiency, our findings on the basis of quantum analysis show the following trends:

(*a*) Patna presses have variations in their speed of work in different sections of jobs.

(*b*) A private press like Tapan Printing Press has the highest speed of work in all its sections, *viz.*, composing, printing and binding. In fact, Tapan Printing Press is more efficient than even some well-organised public presses like the Patna University Press.

(*c*) Patna University Press, being an institutional unit with relatively stronger support, has the lowest speed of work in all its sections. Slow output due to inefficient management, therefore, demands a better attention.

Our enquiry into the sources of manpower supply to printing industry demonstrates that long established tradition of non-apprenticed operatives dominates the other sources of labour supply. This is rather unfortunate. It is, therefore, suggested that the government should take initiative for strict implementation of legal provisions under which presses have to appoint at least one trained personnel in every unit.

An enquiry into the regulations governing the printing industry and their application shows that a sizeable proportion of the presses in the private sector are outside the purview of statutory control. Most of the presses, however, have been registered under the Factories Act. This freedom of operation should contribute to industrial development.

Our analysis of the ownership pattern of printing industry in Patna shows that sole trading form of business organisation is relatively more popular. This further demonstrates the desire of the industry to be slightly independent of legal control.

Comparative Evaluation

Patna is represented by both private and public presses. Although the number of presses under the public sector is few, still it plays a decisive role in the development of printing industry in Patna. One of the main objectives with which the present research project was undertaken has been a comparative evaluation of the working of the presses under the public sector, particularly in the context of the printing industry in Patna. The results of this evaluation are as follows:

(*a*) Government Press, Gulzarbagh, is the first organised printing press in the public sector. The press having been established in 1912 has a grand record of 73 years of printing services to the Government of Bihar and its allied offices. It occupies an important place in the development of printing industry in the State. It fetches 25 per cent profit on outside printing jobs.

(*b*) Our study shows that the general performance of the Government Press has not been encouraging on several fronts. It has some basic problems which hinder its faster growth. The problems are defective structure of organisation, absence of welldefined price policy, inefficient staff, usurping of the Secretariat staff, unscientific and undefined purchasing procedure of machines and materials, irregular printing and publication of government magazines and periodicals, absence of scientific rates of printing, gap in the scale of pay of the press employees and their counterparts in the head office, lack of proper medical care of the employees, insufficient and low quality of printing materials, absence of good industrial relations and other operational or industrial problems.

(*c*) The Government of Bihar has invested about Rs. 100 crores in the press. Therefore, the press owes a responsibility to the Government. The Government is also expected to justify that the press is working to its full satisfaction. Hence, the purpose of such a huge investment in the press has to be measured in terms of its return to the Government. We, therefore, suggest a system of periodical appraisal of performance of the press. It may be done by a State Printing Committee consisting of press authority, government officials, expert of the trade and representatives of Mudrak Parishad. The Committee should present a report of its detailed inquiry to the Government and this will help the Government, Mudrak Parishad and the people of the State to know about the position of the Government Press with regard to the utilisation of its installed capacity and actual output. Moreover, this will be a guideline to others for future.

(*d*) Sound organisational structure and good management are essential pre-requisites to the success of an enterprise. Government Press is no exception to it. The problem of organisation structure and management has, therefore, to be solved in such a manner that it ensures operational efficiency and financial flexibility; and yet does not reduce their accountability to the people and the government. Hence, we suggest that the press section should be granted a separate identity with autonomy and flexibility in its operations. The press should run under the supervision of an independent and competent technical staff, *i.e.*, a printing engineer or an efficient foreman. Thus, both office and workshop of the press should have separate identities to facilitate their smooth functioning as well as promotional reward to the press staff.

(*e*) Budgeting and financial control constitute important tools of industrial management. They help to improve the performance of the printing industry also. Our study, however, reveals some deficiencies with regard to the preparation of budget by the Government Press. Financial transactions of the press are subjected to rigid control of the State Government. The press prepares only the statement of expected expenditure on different items included in the budget which must receive approval of the State Government. The State Government introduces rigidity, and curtails freedom of operation and flexibility necessary for efficient functioning. Moreover, the budget prepared by the press does not include a full list of items. We, therefore, suggest that the press should prepare such a budget which can provide a scientific classification of items of expenditure, and quantum of inputs and outputs. In this way the budget would provide a yardstick for measuring the performance of this press with other public sector presses, *viz.*, private presses.

University Press

Patna University Press has been working since 1951. The pattern of its management before 1980 was entirely different from other presses of the town. It was under the supervision of a Press Management Committee, controlled by an Executive Officer and managed by a supervisor. Thus, the press had several levels of authority for decision making. It was, therefore, difficult to attribute its failures or shortcomings to one single authority. Since 1980, however, the press has been working under the direct supervision of the University with a major change in its management and supervision. But its performance is yet to achieve the desired level of efficiency.

Our investigations show that probably all is not well with the Patna University Press. Whether we talk of output, quality, promptness, soundness or any such achievement, we get a depressing picture. This needs a still more effective management of the press. It is true that a private press may have opportunities which are not easily available to a public press like the University Press. But it is also true that the privileges enjoyed by the University Press are usually rare for a private press. The framework within which the University functions today makes our job slightly difficult in suggesting any specific pattern of management for the University Press. The problem of effective management in the University Press, therefore, needs proper consciousness among the workers, the supervisors and the University administration. To dissolve the Press Management Committee is not a good decision of the University.

Moreover, frequent changes in the pattern of management of the press badly affects staff efficiency.

An attempt was made in this work to conduct time study in the University Press. Even within the constraints of information flow, at least one thing hit us time and again. This was in respect of utilization of duty hours. When workers receive a fixed monthly salary for specified duty hours, they seldom show adequate initiative and efficiency in their work. This is more so when opportunities for further promotion are strictly limited. Thus, elimination of wastage of time should be one of the major considerations of the management of the University Press. If this is done, the rest may not be difficult to achieve. For this purpose a similar study of hours of work of some standard private presses in Patna was also made. The study reveals that the Patna University Press has shorter effective hours. It is, therefore, suggested that for a more productive utilisation of the printing capacity of the press, its working hours should be made to conform to the normal working hours of other presses in Patna.

There is another problem also. The University Press has idle capacity. It is, therefore, suggested that the press should be allowed to accept outside jobs so as to put the idle capacity to profitable use. This needs a suitable system to be devised and developed by the University for handling external assignments of printing. Moreover, there is a big gap between the estimated capacity and the actual out-turn of the press. It is a matter of great concern and steps should be taken to bridge this gap as early as possible. Not only this. The press is still having an old and unscientific schedule of rates for valuation of the out-turn. Hence, the Patna University Press should revise its old rates on more scientific lines and review it every year. But the difficulty is that the press does not maintain an individual out-turn record. It is unproper and unjust because quantity records often show duplicate figures of production. A fair normal allocation of duties of the staff is the only solution to this problem. Attention is also needed to develop an unified system of financial accounting in the press.

The Patna University Press does not prepare Profit and Loss Account and Balance-sheet. Therefore, it is difficult to assess the present position of the press in connection with its financial dealings with the University as well as with outside agencies. We, therefore, suggest that the University should take proper steps in this direction and instruct the press authority to prepare its annual Profit and Loss Account and Balance-sheet as is done by other standard presses of Patna. These accounts should also be properly checked and examined by a duly qualified chartered accountant or an auditor, or the accountant general. A suggested proforma of the Profit and Loss Account and Balance-sheet suitable for adoption by the Patna University Press has been given in the Appendix.

Needless to add that a more economic and efficient working of Patna University Press may be useful to the University as well. This, however, needs attention to the following suggestions:

(*a*) allocation of proper funds for the press;
(*b*) sympathetic attitude of the University authorities towards the press;
(*c*) increased co-operation from the workers;
(*d*) better understanding on the part of the press supervisor;
(*e*) frequent visits to the press by University authorities;
(*f*) periodical check-up of machines; and
(*g*) adequate supply of required materials.

Text Book Press

The Text Book Press is the third press in the public sector covered under the present study. The press is run by autonomous undertaking of the Bihar State Government, *i.e.*, Bihar State Text Book Publishing Corporation. Like other public presses the Text Book Press also has the problems of imperfect organisation structure, improper financial planning, unsound capital structure, poor working capital management, inadequate financial control, absence of skilled personnel, weak industrial relations, poor marketing of products, lower capacity utilization, erratic supply of power, and negligent maintenance of machines. The press is in the clutches of the Corporation and its unscientific control. The Corporation itself does not enjoy the merit of autonomy, flexibility and operational efficiency. In course of our investigation of the Board of Management we found that the Board lacks stability. Powers of the Board are greatly circumscribed and government interference is excessive. The tenure of office of chairman and members of the management is generally undefined. They hold office at the blessing, pleasure or gesture of the State Government. Efficient operation of the Text Book Press, however, depends upon the smoothness and stability of the BSTBPC. But the BSTBPC suffers from political interference and all its related consequences.

The practice of sending government officers on deputation has another limitation. Since their stay in the Corporation is for short duration, they are casual in their approach. Moreover, they rarely develop a feeling of loyalty for and identification of interest with the Corporation. Another disadvantage of the deputation of government bureaucrats is their non-technical skill which proves suicidal for the efficient operation of the press. Hence, we suggest that the top posts be offered to senior employees of the Corporation who have gained the required competence in handling technical and non-technical jobs.

Our study of the capacity utilisation and out-turn of the Text Book Press reveals many loopholes. The annual capacity of out-turn of the press is one crore prints. But the installed capacity has never been achieved. As has been explained earlier the press utilises only about 36 per cent of its capacity because of old and obsolete machines. The rest 64 per cent of the total requirement of printing work is done in other presses. But a number of section heads and other employees of the press workshop reveal another story. The repair works of the machines after breakdown are deliberately delayed and on several occasions proper care is not taken for the maintenance of the printing machines. And this proves a great boon to the authorities. They get 64 per cent of the printing jobs done outside the press which gives them their desired profits. This assiduous practice by the authorities of the Corporation creates a number of other problems which require further investigation. We, therefore, suggest that a time-bound visit of the engineers of G.D.R. Polygraphy Company should be fixed for a regular check-up of the press machines, so that the installed capacity of the machines may be properly utilised and the whimsical attitude is discouraged. The Text Book Press has a rare collection of machines which must be properly maintained. Besides the technical aspects, the financial side of the press too needs control. The imbalance in the capital structure has to be removed. The working capital needs are often ignored. They are required to be met by cash credit arrangements with the banks. This should not be left to casual arrangements.

Some other problems of the Text Book Press include difficulties in collection of bills from wholesellers, marketing of books, power breakdown, etc. Though the Corporation claims to earn profit from publications is unstable and erratic. In fact, the Corporation has sustained losses in the past.

Working Conditions

A statistical review of the working conditions of printing industry in Patna presents a gloomy picture. Most of the private presses work in a smaller space and use inadequate machines. This results in technical incompetence and reduced capacity to work. In the ultimate analysis, such limitations only retard a healthy growth of this industry in future as well.

Our evaluation of the working conditions in printing industry in Patna shows that a number of presses have not been able to make satisfactory provisions for a healthy environment. Besides, a number of presses do not offer adequate incentives to their workers. This is one of the reasons why press workers have a tendency to change their employers frequently. This has resulted in a more reduced efficiency of the industry. Again, inadequate conditions of work have often forced the management either to shift the premises or to close down their business.

A relationship analysis of labour and management in the printing industry in Patna explains how the problem of inadequate personnel arrests the level of efficiency. The press workers too do not stand on a common forum for the protection of their interest. The system of personnel management in the industry is far from satisfactory. This has also been harmful to the operational efficiency of the industry.

While identifying the problems of labour relations we have observed that most of the presses run without proper staff. Either the available personnel are less than the requirement or they do not have adequate experience or training. It is in this state of affairs that even some of the efficient workers often fail to get due recognition. Thus, the printing industry in Patna does not represent defined principles and policies of man power management. Obviously, this further brings down the efficiency of labour as well as management.

It is surprising to find through the present enquiry that a strong association of the printers is not available. It is probably due to this reason that the industry has not been able to receive adequate protection and due participation. Printing industry in Patna, therefore, should organise itself more effectively for the improvement of its conditions.

Quality Control

Our probe into capacity utilization in the printing industry in Patna has shown that about 75 per cent of the presses face shortage of work. This is both due to limitations of job as well as its uneven distribution. This has created a keen competition among the presses of the town. But curiously enough this competition is more between smaller and medium level presses. Quality printers seldom face such competition. At the same time, they do not suffer from inadequacy of work also. This means, therefore, that if the quality of job is improved further, the problem of capacity utilisation may not be as serious as it is today.

Our study demonstrates that a very few presses undertake the printing and publishing of standard books on science, commerce and arts. Publication of books on subjects like medical and engineering has never been made by the presses in this town. Small job works such as cover leaf of Colgate, zarda, match box, candle, vermilion, etc., are done by a few presses. Sometimes novels and pocket books are published but that too by a negligible number of presses. Monthly and weekly magazines are probably a dream for the presses in Patna. Daily newspapers are published by the newspaper presses but not up to standard. Therefore, presses in Patna have a wide scope of printing and publication works in a number of areas such as publication of books, magazines, newspapers, question papers, stationery, envelopes, invitation cards, box making, die stamping, tin printing, metal printing, screen printing, glass, cloth, silk printing, varnishing and waxing, label printing, calendars and diaries and embossing.

But when we talk of quality printing, a number of other problems appear before us. In view of the existing conditions of plant and equipment, premises, personnel and working conditions, quality appears to be a misnomer. Development of this industry, therefore, needs attention on all such problems. Unless the machines are perfect, the workers are fully trained and the environment is healthy, efficiency will not be possible.

Our study shows that there is a multiplicity of printing presses in Patna. They are large in number but small in size. They seldom work together. It is, therefore, suggested that the printers should attempt development of specialization. The size of the presses varies. The size and the nature of the printing job too have an almost similar variety. If presses can specialise in one or two jobs only, they may be able to achieve a better utilization of their capacities resulting in increased profitability. For example, a smaller press may only specialise in printing of letter heads, the other in invitation cards, the third in registers, the fourth in periodicals, the fifth in books, the sixth in newspapers and so on. Of course more than one press may be put under one category. This type of specialisation may facilitate the customers as well and may also contribute to quality development.

State Assistance

Government apathy towards printing industry appears to be unjustified. The government has to demonstrate a fair attitude towards protection and development of the industry. This, therefore, brings into limelight the need for a well-defined government policy on printing industry. Whatever the administrative set-up for granting protection and executive policy at the government level may be, it has to help the workers as well as the management. The group level meets on printing industry organised for the present research project recommended different forms of assistance to this industry by the government, *viz.*, (*a*) fair distribution of printing work, (*b*) appropriate allotment of paper, (*c*) help for financial requirements, (*d*) organised programmes for workers, training, (*e*) standardised rates of printing, (*f*) defined regulations to control the trade operations, and (*g*) collection and publication of periodical data on printing trade. The establishment of a separate department of printing, therefore, is an essential task.

Our analysis of the nature and problems of printing industry in Patna has also highlighted the need for greater efficiency of workers and better co-operation between the printers, workers and the management. In the absence of such efficiency and co-operation the industry is likely to suffer further.

The continuity of an industry also depends upon the economy of its operations. But this economy is not possible unless adequate quantum of resources are available to the industry in time. Moreover, proper costing is another important aspect of financial discipline. In both these directions, however, the prevailing system needs improvement. Unless the printers adopt some sound method of costing their estimates may be misleading. Most of the presses covered under this investigation, however, have not been able to develop adequate resources or reliable methods of costing.

We have now reached the closing part of this research project. An attempt has been made through this investigation to identify the nature and problems of printing industry in Patna. For the purpose of this type of study a large amount of investigations had to be done, studies conducted, process of operations observed, interviews held, group level meets organised, questionnaires distributed, literature studied and visits to almost every press of the town on several occasions had to be made. We are fully conscious of the limitations as well as the gaps of the study. But being an empirical and a pioneer research work, the gaps are almost obvious. The nature of this investigation was such that many presses were not prepared to give us the required information despite repeated requests. Even within these limitations, some efforts have been made to examine briefly the history, organisation, management, labour relations and performance of public and private presses. Let the data, analysis, problems and suggestions contained in the present research serve as a basis to some more fruitful and important studies on this developing industry of the economy.

Appendices

APPENDIX I

Questionnaire on Printing Industry in Patna

A. GENERAL

1. Name of the Press:
2. Location:
   Mohalla:
   Post Office:
   Thana:
   District:
3. Year of Establishment:
4. Registered: Yes/No
5. Registered under:

   (*a*) Factories Act, 1948
   (*b*) Small Scale Industries Act
   (*c*) Shops and Establishment Act
   (*d*) Any other

B. ORGANISATION STRUCTURE

1. Nature of Organisation:

   (*a*) Sole proprietorship
   (*b*) Partnership
   (*c*) Public limited company
   (*d*) Private limited company
   (*e*) Co-operative Society
   (*f*) Any other

2. Building:

(*a*) Ownership:

(*i*) Owned
(*ii*) Rented

(*b*) Area . . . . sq. ft.

3. Procedure of Recording work:

(*a*) Recording System

(*i*) Composing
(*ii*) Printing

C. FINANCIAL SYSTEM

1. Capital Structure:

(*a*) Owned Capital

(*i*) Equity
(*ii*) Preference
(*iii*) Others

(*b*) Borrowed Capital

(*i*) Debenture/Bonds
(*ii*) From Bank
(*iii*) From Government
(*a*) Financial Institutions
(*b*) Government Departments (Industries/Co-operative)
(*iv*) Others.

2. Basis of Payment:

(*a*) Printing: Salary/Daily Wage/Contract
(*b*) Composing: Salary/Daily Wage/Contract
(*c*) Binding: Salary/Daily Wage/Contract
(*d*) Proof Reading: Salary/Daily Wage/Contract

3. Price Policy:

(*a*) Calculation of Cost

(*i*) Materials

(*ii*) Labour
(*iii*) Other Expenses

(*b*) Overheads

(*i*) Profit Margin

4. Budgetary System:

(*a*) Do you prepare any budget? Yes/No
(*b*) If yes, what is the frequency?
Annual/Half-yearly/Quarterly
(*c*) Amount of Allocation: Rs.__________

5. Size of Turnover:

D. LABOUR CONDITIONS

1. Number of Workers:

(*a*) Workshop

(*i*) Composition—Hand/Machines
(*ii*) Machinemen
(*iii*) Fly Boys
(*iv*) Binders
(*v*) Coolies
(*vi*) Others.

(*b*) Office Staff

(*i*) Managing Director
(*ii*) Superintendent
(*iii*) Manager
(*iv*) Accountant
(*v*) Press Assistant
(*vi*) Typists
(*vii*) Peon/Darwan

2. Working Conditions:

(Write yes or no as the case may be)
Is arrangement for the following satisfactory?

(*a*) Health and Safety

(*i*) Lighting
(*ii*) Ventilation

(*iii*) Heating
(*iv*) Cooling
(*v*) Provision for protection from fire

(*b*) Cleanliness

(*i*) Sanitation
(*ii*) Drinking Water
(*iii*) Washing Facilities
(*iv*) Lavatories
(*v*) Wash Basins

(*c*) Welfare Facilities

(*i*) Cycle Shed
(*ii*) Canteen
(*iii*) Rest Room
(*iv*) Dispensaries
(*v*) Hours of Work
(*vi*) Recess

3. Service Conditions:

(*a*) Nature of Employment

(*i*) Permanent
(*ii*) Temporary
(*iii*) Contract
(*iv*) Daily Wages
(*v*) Part Time

(*b*) Main Conditions

(*i*) Leave and Holidays per year

(*a*) Earned Leave
(*b*) Casual Leave
(*c*) Medical Leave
(*d*) Extraordinary Leave
(*e*) Public Holidays

(*ii*) Do you provide housing facilities? If so, in what form?

(*a*) House Allowance
(*b*) Quarters

(*iii*) Penalties and Offences

(*a*) Types of Penalty—Fine/Warning/Wage deduction/Dismissals/Bad Record/Another Punishment

(*c*) Overtime Work

(*i*) Needed or Not
(*ii*) What is the frequency?
(*iii*) Rate of Payment

(*d*) Basis of Promotion—Test/Seniority/Efficiency
(*e*) What is the grievance settlement machinery?

(*i*) Employees Association
(*ii*) Works Committee
(*iii*) Grievance Committee
(*iv*) Industrial Committee
(*v*) Any other macbinery

(*f*) Which of the following facilities are available?

(*i*) Provident Fund
(*ii*) Gratuity
(*iii*) Pension
(*iv*) Worksmen Compensation
(*v*) Bonus
(*vi*) Employees State Insurance
(*vii*) Other Facilities, if any.

(*g*) Which of the following incentive plans are in vogue?

(*i*) Salary increment
(*ii*) *Ad hoc* payment in cash
(*iii*) Prizes
(*iv*) Rewards
(*v*) Promotion
(*vi*) Other incentive, if any

E. TECHNICAL POSITION

1. Printing Machines :

(*a*) Name and Type—Treadle/Flat Bed/Offset/Rotary
(*b*) Size
(*c*) Speed of work

2. Ruling Machines:
   (*a*) Name and Type
   (*b*) Size
   (*c*) Speed of work
3. Stitching Machines:
   (*a*) Type
   (*b*) Size
   (*c*) Speed of work
4. Others:
   (*a*) Type
   (*b*) Size
   (*c*) Speed of work
5. System of Composing:
   (*a*) Hand Composing
   (*b*) Mechanical Type Setting
      (*i*) Mono Composing
      (*ii*) Lino Composing
      (*iii*) Inter Type
      (*iv*) Ludlow
      (*v*) Photo Composing
      (Electronic Composing System)
6. Source of Supply:
   (*a*) Printing Machines
      (*i*) Self-purchase
      (*ii*) Through Government Agencies
      (*iii*) Through Private Agencies
      (*iv*) Others
   (*b*) Metals and Chemicals
      (*i*) Dye Stuff
      (*ii*) Type
   (*c*) Paper Quota
      (*i*) Private Agencies
      (*ii*) Government Agencies

(*iii*) Through Quota System
(*iv*) In the open market

(*d*) Nature and Type of Work

(*i*) Publication of Books—Quantum
(*ii*) Magazines—Monthly/Fortnightly/Daily
(*iii*) Register
(*iv*) Others

7. Trade Policy:

(*a*) Do you feel shortage of work? If so, during which months? What are the factors responsible for such shortage, if any:

(*i*) Limited work
(*ii*) Bad organisation
(*iii*) Shortage of capital
(*iv*) Low quality of work
(*v*) Seasonal fluctuations
(*vi*) Intense compensation
(*vii*) Other factors, if any

(*b*) Problems faced by the Printing Trade in day-to-day operations

(*i*)
(*ii*)
(*iii*)
(*iv*)
(*v*)
(*vi*)
(*vii*)

(*c*) Suggestions for further development of Printing Industry in Patna

(*i*)
(*ii*)
(*iii*)
(*iv*)
(*v*)
(*vi*)

(*d*) Efforts made for Business Expansion

(*i*) Travelling Representatives
(*ii*) Salesmen
(*iii*) Publicity and Advertisement
(*iv*) Personal influence and contacts
(*v*) Others

Signature

APPENDIX II

Particulars of Printing Presses in Patna (as collected from S.D.O. Office)

Sl. No.	Name of Press	Year of Establishment	Location
1	2	3	4
1.	The Searchlight Press	1918	Budh Marg, Patna-4.
2.	Unite Press	1924	Bari Path, Patna-4
3.	Ramesh Printing Press	1927	Mithapur, Patna-1.
4.	Hindustani Press	1933	Bhikhanapahari, Patna-4.
5.	Bharat Printing Works and Label Litho Industry	1935	Patna City Chowk.
6.	Hunkar Press	1938	Budh Marg, Patna-1.
7.	Union Press	1940	Ashok Raj Path, Patna-4.
8.	Sant Mudranalaya	1948	Kadam Kuan, Patna-3.
9.	Tapan Printing Press	1949	Machhuatoli, Patna-4.
10.	Vaishali Press	1951	Annie Besant Road, Patna-4.
11.	Sulekha Printing Press	1959	Kadam Kuan, Patna-3.
12.	Shankar Press	1960	Dariapur Gola, Patna-4.
13.	Azad Hind Press	1960	Arya Kr. Road, Patna-4.
14.	Yatin Press	1962	Langer Toli, Patna-4.
15.	Himachal Press	1962	Kadam Kuan, Patna-4.
16.	Sati Printing Press	1964	Chowk, Patna City.
17.	Art Press	1965	Sultanganj, Patna-6.
18.	Ramayan Press	1965	Lodipur, Patna-1.
19.	Bihar Rajya Shikshak Sangh Press	1966	Exhibition Road, Patna-1.
20.	Bihar Herald Press	1966	Budh Marg, Patna-4.
21.	Bihar Press	1966	Bakerganj, Patna-4.
22.	Chanakya Press	1966	Budh Marg, Patna-4.
23.	City Art Press	1966	Chowk, Patna City.
24.	Dharm Vir Press	1966	Mahendru, Patna-6.
25.	Jan Vani Printing and Publishers	1966	Patna-3.
26.	Kanak Press	1966	Kadam Kuan, Patna-4.
27.	Mohan Press	1966	Gardani Bagh, Patna-1.
28.	Nav Bihar Press	1966	Gardani Bagh, Patna-1.
29.	National Printing House	1966	Kadam Kuan, Patna-4.
30.	Progressive Printing Press	1966	Station Road, Patna.
31.	Print and Printers	1966	Bank Road, Patna-1.
32.	Sanjay Press	1966	Langer Toli, Patna-4.

(*Contd.*)

1	2	3	4
33.	Life Line Press	1966	Kadam Kuan, Patna-4.
34.	Academy Press	1967	Mosallahpur, Patna-6.
35.	Abha Press	1967	Bari Path, Patna-4.
36.	Azad Press	1967	Subzi Bagh, Patna-4.
37.	Bihar Printing Press	1967	Dariapur, Patna-4.
38.	Basant Printing Works	1967	Kadam Kuan, Patna-4.
39.	Everest Press	1967	Prithiwi Raj Path, Patna-3.
40.	Gyan Pith Private Ltd.	1967	Khagawelic Road, Patna.
41.	Label Litho Press	1967	Ramana Road, Patna-4.
42.	Manas Mudrak	1967	Dariapur, Patna-4.
43.	Mahavir Vijoy Press	1967	Rajendra Nagar, Patna.
44.	Muneshwar Art Printing	1967	Exhibition Road, Patna.
45.	Nav Shakti Pvt. Ltd.	1967	Budh Marg, Patna-3.
46.	Prativa Mudranalaya	1967	M.P. Sinha Road, Patna-3.
47.	Prakash Press	1967	Lohanipur, Patna-3.
48.	Rashtra Bhasha Press	1967	Ramna Road, Patna-4.
49.	Sia Ram Press	1967	Tripolia, Patna-7.
50.	Sunil Printing Press	1967	Patna-4.
51.	Sulekha Press	1967	Kadam Kuan, Patna-3.
52.	Shanti Press	1967	Lohanipur, Patna-4.
53.	Vidya Bhushan Press	1967	Mahendru, Patna-6.
54.	Braj Basi Printing Press	1968	Salimpur, Patna-3.
55.	Baidya Nath Press	1968	Bhikhanapahari, Patna-4.
56.	Berojgar Press	1968	Lohanipur, Patna-3.
57.	Janjivan Press	1968	Chiraiatar, Patna-1.
58.	Jagviti Press	1968	Lohanipur, Patna-3.
59.	Jain Press	1968	Machhuatoli, Patna-4.
60.	Karandhar Press	1968	Lodipur, Patna-1.
61.	Kamal Printing Press	1968	Kadam Kuan, Patna-3.
62.	Lakshmi Mudranalaya	1968	Salimpur, Patna-3.
63.	Litho Printing Press	.1968	Subzi Bagh, Patna-4.
64.	Litho Printing Press	1968	Dariapur, Patna-4.
65.	Mail Press	1968	
66.	Print Art	1968	Annie Besant Road, Patna.
67.	Rachna Press	1968	Ranighat, Patna-6.
68.	Reliable Printers	1968	Kadam Kuan, Patna-3.
69.	Raj Press	1968	Patna City.
70.	Uma Shankar Press	1968	Kadam Kuan, Patna-3.
71.	Pioneer Press	1968	Naya Tola, Patna-4.
72.	Novelty Colour Printings	1968	Naya Tola, Patna-4.
73.	Brij Printing Press	1969	Exhibition Road, Patna-1.
74.	B.K. Brother Press	1969	Makhania Kuan, Patna-4.
75.	Chandraprabha Press	1969	Lohanipur, Patna-3.

1	2	3	4
76.	Chitra Press	1969	Ashok Market, Patna-1.
77.	Fine Art Press	1969	Machhuatoli, Patna-4.
78.	Gitanjali Press	1969	Station Road, Patna-1.
79.	Jay Press	1969	Bank Road, Patna-1.
80.	Jyoti Mudranalaya	1969	Chowk, Patna City.
81.	Kant Press	1969	Boring Road, Patna-1.
82.	Lalita Printers	1969	Lohanipur, Patna-4.
83.	Malhotra Press	1969	Pirmuhani, Patna-3.
84.	Niranjan Printing Press	1969	Alamganj, Patna-7.
85.	Navyug Printing Press	1969	Bhikhanapahari, Patna-4.
86.	Parmeshwar Printers	1969	Pirmuhani, Patna-3.
87.	Prasad Printers	1969	Balkishunganj, Patna-7.
88.	Raj Chhapakhana	1969	Jamal Road, Patna-1.
89.	Rakshpal Mudran Kendra	1969	Rajendra Nagar, Patna.
90.	Shri Ram Mudranalaya	1969	Saidpur, Patna-4.
91.	Shramik Mudranalaya	1969	Mahendru, Patna-6.
92.	Sanjay Printing Press	1969	Kadam Kuan, Patna-3.
93.	Shiv Shakti Press	1969	Lodipur, Patna-1.
94.	Shyam Press	1969	Lohanipur, Patna-3.
95.	Shishu Bharat Press	1969	Bhikhanapahari, Patna-4.
96.	Shekhar Colour Printers	1969	Naya Tola, Patna-4.
97.	Samir Printing Press	1969	Lohanipur, Patna-3.
98.	Sharda Press	1969	Guzri Bazar, Patna-7.
99.	Shambhu Press	1969	Rajendra Nagar, Patna.
100.	Vir Mudranalaya	1969	Mithapur, Patna-1.
101.	Veena Press	1969	Bhikhanapahari, Patna-4.
102.	M/s Associate Printers	1970	Station Road, Patna-1.
103.	Bhaijee Press	1970	Meena Bazar, Patna-4.
104.	Bimal Printing Press	1970	Langertoli, Patna-4.
105.	Central Press	1970	Kadam Kuan, Patna-3.
106.	Diamond Printing Works	1970	Naya Tola, Patna-4.
107.	Hindustan Press	1970	Khazanchi Road, Patna-4.
108.	Jan Vani Press	1970	Prithiwipur, Patna-1.
109.	Jayaswal Press	1970	Amaruddin Lane, Patna-3,
110.	Kamal Printing Press	1970	Naya Tola, Patna-4.
111.	Keshri Press	1970	R.K. Avenue, Kadam Kuan, Patna-3.
112.	Kalyan Art Press	1970	Khazanchi Road, Patna-4.

(Contd.)

1	2	3	4
113.	Madhav Press	1970	Dariapur, Patna-4.
114.	Marshal Printers	1970	Kankarbagh, Patna-13.
115.	Murli Dhar Press	1970	Musallahpur, Patna-6.
116.	New Era Printers	1970	G.P. Verma Road, Patna-1.
117.	Patna Times Press	1970	Chirya tar, Patna-1.
118.	Pandey Printing Works	1970	B.M. Das Road, Patna-4.
119.	Rosan Printing Press	1970	Thatheri Bazar, Patna-6.
120.	Rainbow Printing	1970	Nanmuhia, Patna-6.
121.	Shri Sidha Nath Press	1970	Naya Tola, Patna-4.
122.	Shampada Press	1970	Bal Krishnaganj, Patna-7.
123.	Sanjay Press	1970	Khazanchi Road, Patna-4.
124.	Sri Radha Press	1970	Daldali Road, Patna-3.
125.	Satkarita Press	1970	Kadam Kuan, Patna-3.
126.	Uga Tara Printing	1970	Mosallahpur, Patna-6.
127.	Uma Printers	1970	Mosallahpur, Patna-6.
128.	Vishwa Jyoti Press	1970	Naya Tola, Patna-4.
129.	Ashok Press	1971	Mahendru, Patna-6.
130.	Ajay Printers	1971	Patna-3.
131.	Alka Press	1971	Chiryatar, Patna-1.
132.	Anand Printing Press	1971	Chiryatar, Patna-1.
133.	Art & Litho Press	1971	Sultanganj, Patna-6.
134.	Chitra Block and Printing Works	1971	Khazanchi Road, Patna-4.
135.	Chandra Printing Works	1971	Lohanipur, Patna-3.
136.	Ganga Press	1971	Naya Tola, Patna-4.
137.	Himachal Press	1971	Kadam Kuan, Patna-3.
138.	Gyan Pracharak Press	1971	Govind Mitra Road, Patna.
139.	Kumar Bandhu Press	1971	Chandmari Road, Patna-1.
140.	Krishan Press	1971	Lodipur, Patna-1.
141.	Kishor Printing Works	1971	Muradpur, Patna-4.
142.	Lok Mukh Press	1971	Bir Chandra Patel Road, Patna.
143.	Maya Press	1971	Naya Tola, Patna-4.
144.	Miscellaneous Printers	1971	Mithapur, Patna-1.
145.	Manoj Printing Works	1971	Bakarganj, Patna-4.
146.	Nirmal Printing Works	1971	Salimpur, Patna-3.
147.	Navrang Press	1971	Birla Mandir, Patna-4.

1	2	3	4
148.	Pranav Press	1971	Ashok Market, Patna-1.
149.	Prabhakar Printers	1971	Kadam Kuan, Patna-3.
150.	Quality Printing Works	1971	Shahganj, Patna-6.
151.	Raja Press	1971	Kadam Kuan, Patna-4.
152.	Sanjay Printing Press	1971	B.M. Das Road, Patna-4.
153.	Sunil Mudranalaya	1971	Kadam Kuan, Patna-3.
154.	Stylo Printers	1971	Rajendra Path, Patna-1.
155.	Saraswati Press	1971	Jagat Narayan Road, Patna-3.
156.	Vijay Press	1971	Kadam Kuan, Patna-3.
157.	Vijay Printing Press	1971	Rajendra Nagar, Patna.
158.	Janshakti Press	1973	Bir Chand Patel Path, Patna-1.
159.	Aj Press	1981	Frazer Road, Patna-1.
160.	Bir Mudranalaya	1981	Mithapur, Patna-1.

APPENDIX III

Additional List of Printing Presses in Patna (as collected from the former Secretary, Mudrak Parishad)

Sl. No.	Name of Press	Year of Establishment	Location
1	2	3	4
1.	Khadgavilash Press	1880	Bakerganj, Patna-4.
2.	Shyam Bihari Press	1915	Bihari Shaw Lane, Patna-4.
3.	Vani Press	1928	Bhikhana Pahari, Patna-4.
4.	The Indian Nation Press	1930	Frazer Road, Patna-1.
5.	Bharti Press	1935	Mithapur, Patna-1.
6.	Patna Weekly Art Press	1938	Kadam Kuan, Patna-3.
7.	Shri Krishna Press	1938	Diwan Mohalla, Patna-8.
8.	Himalay Press	1939	Govind Mitra Road, Patna-4.
9.	Kalika Press	1941	Arya Kumar Road, Patna-1.
10.	Litho Art Press	1942	Dariyapur, Patna-4.
11.	Nav Nirman Press	1943	Lohanipur, Patna-3.
12.	Prabhat Press	1945	Mithapur, Patna-1.
13.	Kisan Press	1946	Naya Tola, Patna-4.
14.	Ajanta Press	1947	Naya Tola, Patna-4.
15.	Liberty Press	1947	Patna City, Patna-8.
16.	Verma Press	1948	Bakerganj, Patna-4.
17.	Sanjiwan Press	1949	Digha, Patna-1.
18.	Yugantar Press	1951	Ramana Road, Patna-4.
19.	Sarvoday Press	1951	Arya Kumar Road, Patna-1.
20.	Patna University Press	1951	Naya Tola, Patna-4.
21.	Janambhumi Press	1952	Salimpur Ahra, Patna-3.
22.	Bihar Sahyogi Press	1952	Exhibition Road, Patna-1.
23.	Tara Press	1954	Nunmuhia, Patna-7.
24.	Sunil Printing Press	1956	Kazipur, Patna-4.
25.	Janjagran Press	1956	Salimpur Ahra, Patna-3.
26.	Adarsh Press	1957	Arya Kumar Road, Patna-1.
27.	Tarun Bharati Press	1957	B.M. Das Road, Patna-4.
28.	Hind Art Cottage	1958	Govind Mitra Road, Patna-4
29.	Manik Press	1959	Naya Tola, Patna-4.
30.	Vidyarthi Press	1959	Naya Tola, Patna-4.
31.	Dharmayug Press	1960	Jahazi Kothi, Patna-3.
32.	Deomangal Press	1960	Lohanipur, Patna-3.
33.	Hari Printing Press	1960	Kazipur, Patna-4.
34.	Jai Durga Press	1960	Naya Tola, Patna-4.
35.	Arun Art Press	1961	Bhikhanapahari, Patna-4.
36.	Vishnu Yantralay Press	1961	Bhikhanapahari, Patna-4.
37.	Sahni Printing Press	1962	Lohanipur, Patna-3.
38.	Sarat Printing Press	1962	Arya Kumar Road, Patna-1.

1	2	3	4
39.	Yogi Press	1962	Lodipur, Patna-1.
40.	Alok Printing Press	1962	Kazipur, Patna-4.
41.	Atma Press	1962	Boring Road, Patna-1.
42.	Kalpana Press	1962	Mahendru, Patna-6.
43.	Novelty Colour Printing	1962	Naya Tola, Patna-4.
44.	National Block & Printers	1962	Arya Kumar Road, Patna-1.
45.	Bansidhar Press	1963	Langer Toli, Patna-4.
46.	Amar Joyti Press	1963	Hajiganj, Patna-8.
47.	City Art Printers Press	1964	Patna City, Patna-8.
48.	Rai Printing Press	1964	Langer Toli, Patna-4.
49.	Prem Printing Press	1965	Jahazi Kothi, Patna-3.
50.	Rabindra Mudranalaya	1965	Jahazi Kothi, Patna-3.
51.	Sri Tara Printing Press	1965	Ammuddin Lane, Patna-5.
52.	Trimurti Press	1965	Kunkun Singh Lane, Patna-4.
53.	Walmiki Press	1965	Saidpur, Patna-4.
54.	Madhuri Press	1965	Naya Tola, Patna-4.
55.	Nalanda Block Centre	1965	Naya Tola, Patna-4.
56.	Naw Sahani Press	1965	Kazipur, Patna-4.
57.	Hunkar Press	1966	New Market, Patna-1.
58.	Nav Bharat Press	1966	Naya Tola, Patna-4.
59.	Print & Printers	1966	Budh Marg, Patna-1.
60.	Bihar Block Works	1966	Naya Tola, Patna-4.
61.	Anupum Press	1967	Kazipur, Patna-4.
62.	The Art Press	1967	Subzibagh, Patna-4.
63.	Choudhary Art Press	1967	Lohanipur, Patna-3.
64.	Deep Mudranalaya	1967	Mirdha Toli, Patna-7.
65.	Narayan Printers	1967	Khazanchi Road, Patna-4.
66.	Chitra Gupta Press	1968	Makhaniya Kuan, Patna-4.
67.	Bharat Printing Works	1968	Kadam Kuan, Patna-3.
68.	Hindustan Press	1968	Machhuatoli, Patna-4.
69.	Jai Bharat Press	1968	Naya Tola, Patna-4.
70.	Karma Press	1968	Lodipur, Patna-1.
71.	M.F.J. Press	1968	Kankarbagh, Patna-20.
72.	Jyoti Press	1968	Budh Marg, Patna-1.
73.	New Standard Printing Press	1970	Naya Tola, Patna-4.
74.	Magadh Press	1970	Sultanganj, Patna-6.
75.	Datta Printing Press	1970	Kadam Kuan, Patna-3.
76.	Shri Gyan Press	1970	Patna
77.	Radha Press	1971	Kadam Kuan, Patna-3.
78.	Saraswati Press	1971	Kadam Kuan, Patna-3.
79.	Calcutta Art Press	1971	Dariapur, Patna-4.
80.	Arun Printing Press	1971	Kunkun Singh Lane, Patna-4.

(*Contd.*)

1	2	3	4
81.	Anand Press	1971	Bakerganj, Patna-4.
82.	Chandradaya Press	1972	Saidpur, Patna-4.
83.	Arun Printing Press	1972	Sidhsumeru Nagar, Patna-7.
84.	Ajit Printing Press	1972	Langer Toli, Patna-4.
85.	Basant Printing Press	1972	Exhibition Road, Patna-1.
86.	City Press	1972	City Court, Patna-8.
87.	Daya Printing & Publishing	1972	Mahendru, Patna-6.
88.	International Printing Works	1973	Lohanipur, Patna-3.
89.	Modern Press	1973	Kazipur, Patna-4.
90.	Prabudh Press	1973	Mahendru, Patna-6.
91.	Rana Printing Works	1973	Patna
92.	Shankar Press	1973	Saidpur, Patna-4.
93.	Sada-e-Aam Press	1973	Subzibagh, Patna-4.
94.	Shyam Mudranalaya	1973	Rajendra Nagar, Patna.
95.	Anjuman Press	1974	Machhuatoli, Patna-4.
96.	Bishwa Bandhu Press	1974	Station Road, Patna-1
97.	Bubna Press	1974	Chowk, Patna-8.
98.	Himalay Press	1974	Mahendru, Patna-6.
99.	Law Press	1975	Jamal Road, Patna-1.
100.	Patna Times Press	1975	Lohia Nagar, Patna-10.
101.	Jyoti Prakashan Press	1975	Patna Gaya Road, Patna-1.
102.	Splinter	1975	Patna City, Patna-8.
103.	Maithili Prakashan Press	1975	Naya Tola, Patna-4.
104.	Nav Jiwan Press	1976	Digha, Patna-1.
105.	Nav Bihar Printing Press	1976	Nala Road, Patna-3,
106.	Bhasker Press	1976	Subzibagh, Patna-4.
107.	Sardar Printing Press	1977	Bal Kishunganj, Patna-7.
108.	Sidharth Press	1977	Naya Tola, Patna-4.
109.	Agrawal Press Works	1977	Patna City.
110.	Sarda Printers	1977	Patna City.
111.	Jagdish Printing	1978	City Court, Patna-8.
112.	Ajanta Printers	1978	Patna City.
113.	Shankar Art Press	1978	Machharhatta, Patna-8.
114.	Patel Printers	1978	Chhoti Patan Devi, Patna-7.
115.	Sujata Press	1979	Maccharhatta, Patna-8.
116.	Saraswati Press	1979	Kachauri Gali, Patna-8.
117.	Lakshmi Printers	1979	Mangal Talab, Patna-8.
118.	Quality Printers	1979	Kali Asthan, Patna-8.
119.	Naya Printers	1980	Patna City.
120.	National Press	1980	Phulouri Gali, Patna-8.
121.	Calendar Press	1980	Hajiganj, Patna-8.
122.	Naw Bharat Printing Press	1980	Patna City, Patna-8.

1	2	3	4
123.	Rekha Printing Press	1981	Gurhatta, Patna-8.
124.	Nutan Printing Press	1981	Musallahpur, Patna-6.
125.	Mithila Box and Colouring Works	1981	Musallahpur, Patna-6.
126.	Luna Printing Works Press	1981	Musallahpur, Patna-6.
127.	Maa Durga Press	1982	Musallahpur, Patna-6.
128.	Sri Ram Printer	1982	Mahendru, Patna-6.
129.	J.T.S. Press	1982	Dariyapur, Patna-4.
130.	Lakhan Press	1982	B.M. Das Road, Patna-4.
131.	Krishna Printing Press	1982	Makhaniya Kuan, Patna-4.
132.	Sobha Press	1982	Lohia Nagar, Patna-10.
133.	Mahavir Press	1982	Musallahpur, Patna-6.
134.	J.K. Offset	1983	Musallahpur, Patna-6.
135.	Janhit Press	1983	Musallahpur, Patna-6.
136.	Kamal Printing Press	1983	Musallahpur, Patna-6.
137.	Chaturbhuj Printing	1983	Tikia Toli, Patna-6.
138.	Ashutosh Printing Press	1983	Ranighat, Patna-6.
139.	Sinha Printing Press	1983	Naya Tola, Patna-4.
140.	Ajay Press	1983	Langer Toli, Patna-4.
141.	Bihar Journal Press	1983	Kadam Kuan, Patna-3.
142.	Baishno Press	1983	Patna City.
143.	Trinetra Printing Press	1983	Naya Tola, Patna-4.
144.	Awadh Printers	1984	Langer Toli, Patna-4.
145.	Sunder Press	1984	Balkishunganj, Patna-7.
146.	Patliputra Press	1984	Kadam Kuan, Patna-3.

APPENDIX IV

Supplementary List of Printing Presses in Patna (as collected from the present Secretary, Mudrak Parishad)

Sl. No.	Name of Press	Year of Establishment	Location
1	2	3	4
1.	Satya Sudhakar Press	1900	Chowk, Patna City
2.	Government Press	1912	Gulzarbagh, Patna-7.
3.	Burki Art Press	1915	Subzibagh, Patna-4.
4.	District Board Press	1919	Patna-4.
5.	Hind Press	1922	Govind Mitra Road, Patna-4.
6.	Jagdish Printing Press	1928	City Court, Patna-7.
7.	University Press	1943	Jahazi Kothi, Patna-3.
8.	Spark Press	1944	Pirmohani, Patna-3.
9.	Janta Press	1948	Naya Tola, Patna-4.
10.	Jay Hind Press	1952	Langer Toli, Patna-4.
11.	Bihar Co-operative Societies Press Ltd.	1954	Lohanipur, Patna-3.
12.	Durga Press	1954	Bhattacharya Road, Patna-1.
13.	Magadh Mudranalaya	1955	R.K. Avenue, Patna-3.
14.	Sila Printing Press	1957	Kajipur, Patna-4.
15.	Gyan Press	1957	Machhuatoli, Patna-4.
16.	Liberty Press	1958	Bakerganj, Patna-4.
17.	Jiwan Jyoti Press	1959	Hajiganj, Patna-8.
18.	Arbind Press	1960	Balkishunganj, Patna-7.
19.	Bhola Printing Press	1960	Mahendru, Patna-6.
20.	Ghan Shyam Press	1960	Navin Kothi, Patna-4.
21.	Pustakalaya Press	1960	Tripolia, Patna-7.
22.	Sarv Seva Press	1960	Kadam Kuan, Patna-3.
23.	Press Republic	1961	Naya Tola, Patna-4.
24.	Pandey Printing Press	1961	Patna.
25.	Bareria Art Press	1962	Patna
26.	Gyanand Press	1962	Nunmuhia, Patna-7.
27.	Prakash Press	1962	Kajipur, Patna-4.
28.	Sanand Press	1962	Lohanipur, Patna-3.
29.	Moonlight Press	1963	Kadam Kuan, Patna-3.
30.	Mahendra Printing Press	1963	City Court, Patna-7.
31.	Goodwill Brothers Printers	1963	Kadam Kuan, Patna-3.
32.	Gaya Art Press	1963	Kadam Kuan, Patna-3.

1	2	3	4
33.	Golden Printing Press	1964	Mangal Talab, Patna-8.
34.	Manchand Press	1964	Digha, Patna-11.
35.	Vinod Press	1965	Chandmari Road, Patna-16.
36.	Lakshmi Press	1965	Bakerganj, Patna-4.
37.	Abhuday Press	1966	New Market, Patna-1.
38.	Barman Press	1966	R.K. Avenue, Patna-3.
39.	Gantantra Press	1966	Kadam Kuan, Patna-3.
40.	Indian Press	1966	Balkrishunganj, Patna-7.
41.	Kumar Press	1966	Abulash Lane, Patna-4.
42.	Karm Yogi Press	1966	Arya Kumar Road, Patna-4.
43.	Navin Printing Press	1966	Gayaghat, Patna-7.
44.	Shushila Printing Press	1967	Bhikhanapahari, Patna-4.
45.	Shri Chaudhary Press	1967	Tikeya Toli, Patna-6.
46.	Indian Light Press	1967	Dariapur, Patna-4.
47.	Hanuman Press	1967	Devendranath Lane, Patna-4.
48.	Agrawal Press	1967	Mithapur, Patna-1.
49.	Arunoday Press	1967	Mahendru, Patna-6.
50.	D.K. Printers and Stationers	1968	R.K. Avenue, Patna-3.
51.	Liberal Press	1968	Station Road, Patna-1.
52.	Asha Printers	1969	Abulash Lane, Patna-4.
53.	Bhola Art Printers	1969	Gulzarbagh, Patna-7.
54.	Appolo Printers	1970	Naya Tola, Patna-4.
55.	Vinay Art Press	1970	Tripolia, Patna-7.
56.	Suryoday Press	1970	Musallahpur, Patna-6.
57.	Patliputra Press	1970	Langer Toli, Patna-4.
58.	Patna Hind Press	1970	Sultanganj, Patna-6.
59.	Damodar Printing Press	1970	Lohanipur, Patna-3.
60.	Brinda Printing Press	1970	Hajiganj, Patna-8.
61.	Alka Press	1970	Patna City, Patna-8.
62.	Akshar Press	1970	Orawin Kothi, Patna-4.
63.	Sangh Mudranalaya	1971	Rajendra Nagar, Patna.
64.	Sunil Mudranalaya	1971	Kishan Cold Storage, Patna-6.
65.	Janak Kalyan Press	1971	Naya Tola, Patna-4.
66.	Singh and Sinha Press	1972	Kadam Kuan, Patna-3.
67.	Standard Press	1972	Bhikhanapahari, Patna-4.
68.	Rose Printer	1972	Kadam Kuan, Patna-3.
69.	Gupteshwar	1972	B.M. Das Road, Patna-4.
70.	Harinam Press	1972	Jhauganj, Patna-8.
71.	Bihar State Text Book Corporation Press	1972	Budh Marg, Patna-1.

(*Contd.*)

1	2	3	4
72.	Amar Bharti Press	1973	Patna City, Patna-8.
73.	Shri Girja Press	1973	Kadam Kuan, Patna-3.
74.	Amarbind Press		Dondi Bazar, Patna-8.
75.	Ajay Art Printers		Mahendru, Patna-6.
76.	Bharat Mail Press		Hajiganj, Patna-8.
77.	Bihar Provincial Press		Kadam Kuan, Patna-3.
78.	Bharat Bhushan Press		Patna-1.
79.	Bihar Envelop Press		Dariapur, Patna-4.
80.	Bombay Envelop Press		Dariapur, Patna-4.
81.	Hindu Press		Sultanganj, Patna-6.
82.	Jai Shree Art Cottage		Bhikhanapahari, Patna-4.
83.	Janbharti Press		Kadam Kuan, Patna-3.
84.	Karmshil Art Press		B.M. Das Road, Patna-4.
85.	Lipika Press		Govind Mitra Road, Patna-4.
86.	Modern Printers		Langer Toli, Patna-4.
87.	Modi Press		Machhuatoli, Patna-4
88.	Nayan Bharti Press		Hajiganj, Patna-8.
89.	Nawjiwan Press		Bhikhanapahari, Patna-4.
90.	New Printing Press		Langer Toli, Patna-4.
91.	Prakash Printing Press		Pirmuhani, Patna-3.
92.	Prachi Press		Kadam Kuan, Patna-3.
93.	Pratibha Press		New Market, Patna-1.
94.	Pushpanjali Press		Lohanipur, Patna-3.
95.	Ram Krishna Press		Sultanganj, Patna-6.
96.	Rupkala Press		Exhibition Road, Patna-1.
97.	R.N.C. Press		R.K. Avenue, Patna-3.
98.	Shobha Press		Station Road, Patna-1.
99.	Sima Press		Station Road, Patna-1.
100.	Swastika Printing Press		Hajiganj, Patna-8.
101.	Smart Press		Mahendru, Patna-6.
102.	Standard Printing Press		Dariapur, Patna-4.
103.	Samajseva Mudranalaya		Lohanipur, Patna-3.
104.	Surendra Mudranalaya		Nawal Ashram, Patna-3.
105.	Universal Press		Mahendru, Patna-6.
106.	Vidyasagar Mudranalaya		Lohanipur, Patna-3.
107.	Yugvani Press		Bakerganj, Patna-4.
108.	Rup Rekha Press		Station Road, Patna-1.

APPENDIX V

Presses under Pilot Survey

Sl. No.	Name of presses	Year of Establishment	Location
1.	Ajanta Press	1947	Naya Tola, Patna-4.
2.	Ashok Press	1947	Mahendru, Patna-6.
3.	Amarjyoti Press	1963	Hajiganj, Patna-8.
4.	Bihar Block Works	1966	Naya Tola, Patna-4.
5.	Chandroday Press	1972	Saidpur, Patna-4.
6.	Chanda Printing Works	1972	Kadam Kuan, Patna-3.
7.	City Art Printers	1966	Patna City, Patna-8.
8.	Gyanpith Printing Ltd.	1950	Khazanchi Road, Patna-4.
9.	Hindustani Press	1933	Ramna Bagh, Patna-4.
10.	Jaya Durga Press	1960	Naya Tola, Patna-4.
11.	Janambhumi Press	1952	Patna City, Patna-8.
12.	Kalika Press	1941	Arya Kumar Road, Patna-1.
13.	Khadgavilas Press	1880	Bakerganj, Patna-4.
14.	Label Litho Press	1950	Ramna Road, Patna-4.
15.	Liberty Press	1947	Guzri, Patna-8.
16.	Narayan Printers	1967	Khazanchi Road, Patna-4.
17.	Navnirman Press	1943	Kadam Kuan, Patna-3.
18.	Ramesh Printing Press	1927	Mithapur, Patna-1.
19.	Sri Sidhnath Press	1970	Naya Tola, Patna-4.
20.	Sanjiwan Press	1949	Digha, Patna-1.
21.	Tapan Printing Press	1949	Arya Kumar Road, Patna-1.
22.	Tara Press	1954	Nunmuhia, Patna-7.
23.	Tarun Bharat Press	1957	B.M. Das Road, Patna-4.
24.	The Searchlight Press	1918	Budh Marg, Patna-1.
25.	United Press	1924	Bari Road, Patna-4.
26.	Vaishali Press	1951	Annie Besant Road, Patna-4.
27.	Vishwajyoti Press	1970	Naya Tola, Patna-4.
28.	Yugantar Press	1951	Ramna Bagh, Patna-4

APPENDIX VI

Presses taken for General Survey

Sl. No.	Name of press	Year of establishment	Location
1	2	3	4
1.	Academy Press	1966	Musallahpur, Patna-4.
2.	Anupum Printing Works	1967	Kazipur, Patna-4.
3.	Arun Art Printers	1961	Bhikhanapahari, Patna-4.
4.	Anand Printers	1971	Bakerganj, Patna-4.
5.	Adarsh Press	1957	Arya Kumar Road, Patna-1.
6.	Atma Press	1962	Boring Road, Patna-1.
7.	Art Press	1965	Mahendru Patna-6.
8.	Ajay Printers	1970	Chiraiatar, Patna-1.
9.	Abha Press	1967	Machhuatoli, Patna-4.
10.	Art & Litho Press	1971	Sultanganj, Patna-6.
11.	Ajay Printers	1971	Kadam Kuan, Patna-3.
12.	Anand Printing Press	1971	Chiraiatar, Patna-1.
13.	Azimabad Press	1973	Subzi Bagh, Patna-4.
14.	Bansidhar Press	1963	Langer Toli, Patna-4.
15.	Bihar Printing Press	1967	Dariapur, Patna-4.
16.	Bihar Rajya Shikshak Sahyog Sangh Press	1966	Exhibition Road, Patna-1.
17.	Bharat Printing Works	1968	Kadam Kuan, Patna-3.
18.	Bimal Printing Works	1970	Langertoli, Patna-4.
19.	Bihar Press	1966	Rupak Cinema, Patna-4.
20.	Berojgar Press	1968	Lohanipur, Patna-3.
21.	Bir Mudranalaya	1981	Mithapur, Patna-1.
22.	Chitra Block Printing Works	1971	Naya Tola, Patna-4.
23.	Chanakya Press	1966	New Market, Patna-1.
24.	Chitragupta Press	1968	Makhania Kuan, Patna-4.
25.	Chandra Prabha Press	1969	Lohanipur, Patna-3.
26.	Central Press	1970	Kadam Kuan, Patna-3.
27.	Dharamyug Press	1960	Kadam Kuan, Patna-3.
28.	Dip Mudranalaya	1967	Alamganj, Patna-7.
29.	Dharamvir Press	1966	Mahendru, Patna-6.
30.	Dayal Printing Press	1972	
31.	Fine Art Press	1967	Machhuatoli, Patna-4.
32.	Gitanjali Press	1969	Station Road, Patna-1.
33.	Gyanoday Press	1965	Govind Mitra Road, Patna-4.
34.	Himachal Press	1962	Kadam Kuan, Patna-3.
35.	Hunkar Press	1966	New Market, Patna-1.

1	2	3	4
36.	Hind Art Cottage	1957	Govind Mitra Road, Patna-4.
37.	Hindustan Press	1912	Khazanchi Road, Patna-4.
38.	Hunkar Printing Works	1936	Budh Marg, Patna-1.
39.	Janjagaran Press	1956	Salimpur, Patna-3.
40.	Jyoti Mudranalaya	1969	Dundi Bazar, Patna-8.
41.	Janvani Press	1970	Patna-3.
42.	Jagriti Press	1968	Lohanipur, Patna-3.
43.	Jay Bharat Press	1966	Naya Tola, Patna-4.
44.	Kanak Press	1966	Kadam Kuan, Patna-3.
45.	Kalyan Art Press	1970	Khazanchi Road, Patna-4.
46.	Kisan Press	1947	Naya Tola, Patna-4.
47.	Kalpana Press	1962	Musallahpur, Patna-6.
48.	Kant Press	1968	Boring Road, Patna-1.
49.	Krishna Press	1968	Lodipur, Patna-1.
50.	Kishor Printing Works	1962	Muradpur, Patna-4.
51.	Lokmukh Press	1971	Vir Chand Patel Road, Patna-1.
52.	Litho Printing Press	1968	Subzi Bagh, Patna-4.
53.	Litho Art Press	1942	Langer Toli, Patna-4.
54.	Madhuri Press	1965	Naya Tola, Patna-4.
55.	Manik Press	1954	Naya Tola, Patna-4.
56.	Madhav Press	1970	Dariapur, Patna-4.
57.	Miscellaneous Press	1971	Mithapur, Patna-1.
58.	Marshal Printers	1970	Kankarbagh, Patna-16.
59.	M.F.J. Press	1968	Kankarbagh, Patna-16.
60.	Manoj Printing Works	1971	Subzi Bagh, Patna-4.
61.	Muneshwar Printing Press	1967	Exhibition Road, Patna-1.
62.	New Standard Printing	1970	Naya Tola, Patna-4.
63.	Niranjan Printing Press	1969	Alamganj, Patna-7.
64.	National Printing House	1966	Kadam Kuan, Patna-3.
65.	Nalanda Block Centre	1965	Naya Tola, Patna-4.
66.	New Sahani Printing Press	1965	Kazipur, Patna-4.
67.	New Era Printers	1970	S.P. Verma Road, Patna-1.
68.	Navrang Press	1971	Bhawar Pokhar, Patna-3.
69.	Novelty Colour Printing Works	1962	Naya Tola, Patna-4.
70.	Prabudh Press	1967	Mahendru, Patna-6.
71.	Prakash Press	1969	Lohanipur, Patna-3
72.	Prem Printing Press	1965	Arya Kumar Road, Patna-4.
73.	Prabhat Press	1945	Mithapur, Patna-1.

(*Contd.*)

1	2	3	4
74.	Print and Printers	1967	Lodipur, Patna-1.
75.	Parmeshwar Printer	1965	Pirmohani, Patna-3.
76.	Pranav Press	1970	Ashok Market, Patna-1.
77.	Prabhakar Printers	1971	M.P. Sinha Road, Patna-3.
78.	Progressive Printing Press	1966	Station Road, Patna-1.
79.	Pioneer Press	1968	Naya Tola, Patna-4.
80.	Rachana Press	1968	Musallahpur, Patna-6.
81.	Raj Press	1968	Salimpur Ahra, Patna-3.
82.	Rakshpal Mudran Kendra	1969	R.K. Avenue, Patna-3.
83.	Rainbow Printing Press	1969	Nunmuhia, Patna-7.
84.	Ravindra Mudranalaya	1965	Jahazi Kothi, Patna-3.
85.	Radha Press	1958	Daldali Road, Patna-3.
86.	Roy Printing Works	1964	Langertoli, Patna-4.
87.	Ramayan Press	1965	Golghar, Patna-1.
88.	Sunil Printing Press	1956	Kazipur, Patna-4.
89.	Sulekha Printing Works	1959	Kadam Kuan, Patna-3.
90.	Shri Shyam Bihari Press	1915	Bihari Shaw Lane, Patna-4.
91.	Sharat Printing Press	1962	Arya Kumar Road, Patna-4.
92.	Shri Krishna Press	1939	Diwan Mohalla, Patna-8.
93.	Shati Printing Press	1964	Jhauganj, Patna-8.
94.	Sanjay Printing Press	1969	B.M. Das Road, Patna-4.
95.	Shyam Press	1969	Lohanipur, Patna-3.
96.	Sant Mudranalaya	1948	Kadam Kuan, Patna-3.
97.	Shankar Press	1960	Dariapur, Patna-3.
98.	Shyam Printing & Label Industry	1967	Jhauganj, Patna-8.
99.	Shanti Press	1967	Lohanipur, Patna-3.
100.	Stylo Press	1971	Rajendra Path, Patna-1.
101.	Sampada Press	1970	Balkishunganj Patna-7.
102.	Shekhar Colour Printing Press	1969	Naya Tola, Patna-4.
103.	Sarda Press	1967	Guzari Bazar, Patna-8.
104.	Siaram Press	1967	Gulzarbagh, Patna-7.
105.	Sanjay Printing Press	1971	B.M. Das Road, Patna-4.
106.	Shri Radha Press	1970	Daldali Road, Patna-3.
107.	Uma Printers	1970	Musallahpur, Patna-6.
108.	Umashankar Press	1968	Kadam Kuan, Patna-3.
109.	Vani Press	1928	Bhikhanapahari, Patna-4.
110.	Vir Mudranalaya	1969	Mithapur, Patna-1.
111.	Vidya Bhushan Press	1967	Mahendru, Patna-6.
112.	Verma Press	1949	Bakerganj, Patna-4.
113.	Vijay Printing Press	1971	Rajendra Nagar, Patna.

APPENDIX VII

Presses Responding to the Questionnaire

Sl. No.	Name of Press	Year of establishment	Owner	Location
1	2	3	4	5
1.	Academy Press	1966	Rameshwar Pd.	Mussallahpur, Patna-6.
2.	Anupam Printing Press	1967	Shanker Pd. Ghose	Kazipur, Patna-4.
3.	Ajanta Press	1947	Jai Nath Mishra	Naya Tola, Patna-4.
4.	Ashok Press	1947	Sivaji Sinha	Mahendru, Patna-6.
5.	Amar Jyoti Press	1963	Prabhat Kr. Sinha	Hajiganj, Patna-8.
6.	Arun Art Press	1961	Arun Prasad	Bhikhanapahari, Patna-4.
7.	Anand Printers	1971	Sita Ram Chaurasia	Bakerganj, Patna-4.
8.	Adarsh Press	1957	Birendra Kumar	Arya Kr. Road, Patna-1.
9.	Atma Press	1962	Atma Singh	Boring Road, Patna-1.
10.	Art Press	1965	Md. Mustafa	Sultanganj, Patna-6.
11.	Ajay Printers	1970	Braj Mohan Sahay	Chiraiatar, Patna-1.
12.	Bansidhar Press	1963	Bansidhar Singh	Langertoli, Patna-4.
13.	Bihar Printing Press	1967	Madan Mohan Singh	Dariyapur, Patna-4.
14.	Bihar Rajya Sikshak Sangh Press	1966	Private Ltd.	Exhibition Road, Patna-1.
15.	Bharat Printing Press	1968	Bharat Singh	Kadam Kuan, Patna-3.
16.	Bihar Block Works	1966	Jitendra Nath Sirkar	Naya Tola, Patna-4.
17.	Chanakya Press	1966	Bindeshwari Thakur	New Market, Patna-1.

(Contd.)

APPENDIX VII (*Contd.*)

1	2	3	4	5
18.	Chitra Block and Printing Works	1971	Arun Kr. Jha	Naya Tola, Patna-4.
19.	Chandroday Press	1972	Chandreshwar Pd.	Saidpur, Patna-4.
20.	Chanda Printing Works	1971	Lal Bahadur Pd.	Lohanipur, Patna-3.
21.	City Art Press	1966	Om Prakash	Chowk, Patna-8.
22.	Dharamyug Press	1960	Jagdish Jha	Alamganj, Patna-7.
23.	Dip Mudranalaya	1967	Dip Singh	Alamganj, Patna-7.
24.	Fine Art Press	1967	Vishwa Mohan Kr. Singh	Machhuatoli, Patna-4.
25.	Gyanpith Printing Ltd. Press	1950	Tripershan Pandey	Khazanchi Road, Patna-4.
26.	Gyanoday Press	1965	Basisth Narain Chaudhary	Govind Mitra Road, Patna-4.
27.	Hindustani Press	1933	Devendra Kr. Mishra	Naya Tola, Patna-4.
28.	Himachal Press	1962	Rajni Kant Sinha	New Market, Patna-1.
29.	Hind Art Cottage	1957	Jagdish Shankar	Govind Mitra Road, Patna-4.
30.	Hunkar Press	1938	Umesh Prasad	Budh Marg, Patna-1.
31.	Janjagran Press	1956	Surendra Nath	Salimpur, Patna-3.
32.	Jyoti Mudranalaya	1969	Dharamdeo Prasad	Dundi Bazar, Patna-8.
33.	Jay Durga Press	1960	Nagendra Pd. Singh	Naya Tola, Patna-4.
34.	Janambhumi Press	1952	Girdhar Lal Gerg	Chowk, Patna-8.
35.	Jay Bharat Press	1966	Bansi Singh	Naya Tola, Patna-4.
36.	Kalika Press	1941	Brahmdeo Pd.	Arya Kr. Road, Patna-1.
37.	Khadgavilas Press	1880	Sharandhar Pd. Singh	Bakerganj, Patna-4.
38.	Kisan Press	1947	Hansdeo Mehta	Naya Tola, Patna-4.
39.	Kalpana Press	1962	Guneshwar Prasad	Musallahpur, Patna-6.
40.	Kant Press	1968	Shyama Kant Sinha	Boring Road, Patna-1.
41.	Krishna Press	1968	Jogendra Pd. Singh	Lodipur, Patna-1.

42.	Kishor Printing Press	1962	Uday Narayan Sinha	Musallahpur, Patna-6.
43.	Litho Art Press	1942	Md. Yasin	Langertoli, Patna-4.
44.	Label Litho Press	1950	Sin Muzaffarpuri	Ramna Road, Patna-4.
45.	Liberty Press	1947	Abhay Mohan	Guzri, Patna-8.
46.	Madhuri Press	1965	Din Dayal Pd.	Naya Tola, Patna-4.
47.	Manik Press	1970	Raj Nath Press	Dariyapur, Patna-4.
48.	Madhav Press	1954	Kishor Pd. Singh	Naya Tola, Patna-4.
49.	Miscellaneous Press	1971	Vinod Bihari Agrawal	Mithapur, Patna-1.
50.	Nalanda Block Centre	1965	Kamla Pd. Verma	Naya Tola, Patna-4.
51.	New Sahni Printing Press	1965	Ram Balak Sahani	Kazipur, Patna-4.
52.	New Era Printers	1970	Syed Welayat Ali	S.P. Verma Road, Patna-1.
53.	Navrang Press	1971	Brajnandan Nirmal	Bhawar Pokhar, Patna.
54.	Novelty Colour Printing Works	1962	Jai Nath Mishra	Naya Tola, Patna-4.
55.	Narayan Printers	1967	Shambhu Sharan Srivastava	Khazanchi Road, Patna-4
56.	Naw Nirman Press	1943	Mahavir Prasad	Lohanipur, Patna-3.
57.	Prabhat Press	1945	Prabhat Sarkar	Mithapur, Patna-1.
58.	Print and Printers	1967	Mrs. Sharla Rohatgi	Lodipur, Patna-1.
59.	Parmeshwar Printers	1969	Mrs. Jiwan Kapoor	Pirmohani, Patna-3.
60.	Pranav Press	1970	Pranav Kumar	Ashok Market, Patna-1.
61.	Prabhakar Printers	1971	Prem Lal	M.P. Sinha Road, Patna-3.
62.	Progress Printing Press	1966	Anand Mohan Banerjee	Station Road, Patna-1.

(*Contd.*)

APPENDIX VII—(*Contd.*)

1	2	3	4	5
63.	Pioneer Press	1968	Ravindra Prasad	Naya Tola, Patna-4.
64.	Ravindra Mudranalaya	1965	Smt. Mirdula Singh	Jahaji Kothi, Patna-3.
65.	Ramesh Printing Press	1927	Narendra Bakshi	Mithapur, Patna-1.
66.	Radha Press	1958	Shridhar Das	Daldali Road, Patna-3.
67.	Roy Printers Works	1964	Bimal Kumar Roy	Langertoli, Patna-4.
68.	Ramayan Press	1965	Santos Kr. Singh	Golghar, Patna-1.
69.	Sri Sidhnath Press	1970	Sidh Nath Pd. Roy	Naya Tola, Patna-4.
70.	Sanjivan Press	1949	Run by Mission	Dargah, Patna.
71.	Sunil Printing Press	1956	Ananda Pd. Singh	Kazipur, Patna-4.
72.	Sulekha Printing Works	1959	Mahadeo Chand	Kadam Kuan, Patna-3.
73.	Shri Shyam Bihari Press	1915	Shyam Bihari Pd.	Arya Kr. Road, Patna-1.
74.	Sharat Printing Press	1962	Sharat Kumar	Bilan Sahu Lane, Patna-4.
75.	Sri Krishna Press	1930	Krishna Prasad	Diwan Mohalla, Patna-8.
76.	Sati Printing Press	1964	Gaurishankar Jhunjhunwala	Phulauri Gali, Patna-8.
77.	Shyam Press	1969	Hari Pd. Sinha	Lohanipur, Patna-3.
78.	Sant Mudranalaya	1948	K.A. Figg	Kadam Kuan, Patna-3.
79.	Shankar Press	1960	Sri Radha Krishna Singh	Dariyapur, Patna-3.
80.	Sanjay Printers	1971	Baidya Nath Pd.	B.M. Das Road, Patna-4.
81.	Shyam Printer and Label Industry	1967	Shyam Sundar Pd.	Jhauganj, Patna-8.
82.	Tapan Printing Press	1949	M.M. Bose	Arya Kr. Road, Patna-1.
83.	Tara Press	1954	Vishwa Nath Maitin	Nanmuhiya, Patna-4.

84.	Tarun Bharat Printing	1957	Asharfi Roy Sharma	B.M. Das Road, Patna-4.
85.	The Searchlight Press	1918	Private Ltd.	Budh Marg, Patna-1.
86.	United Press	1924	Private Ltd.	Hathwa Market, Patna-4.
87.	Uma Printers	1970	Umashankar Srivastava	Musallahpur, Patna-6.
88.	Vaishali Press	1951	Hans Raj	Annie Besant Road, Patna-4.
89.	Vishwa Joyti Press	1970	Vishwa Nath Jha	Naya Tola, Patna-4.
90.	Verma Press	1949	Ram Pd. Verma	Bakerganj, Patna-4.
91.	Vijay Printing Press	1971	Shyam Kr. Prasad	Rajendra Nagar, Patna.
92.	Yugantar Press	1951	Devendra Nath Mishra	Ramna Bagh, Patna-4.

APPENDIX VIII

Presses not Responding to the Questionnaire

Sl. No.	Name of Press	Location
1	2	3
	1. Untraceable Presses	
1.	Abha Press	Machhuatoli, Patna-4
2.	Art and Litho Press	Sultanganj, Patna-6
3.	Ajay Printers	Kadam Kuan, Patna-3
4.	Anand Printing Press	Chiraiatar, Patna-1
5.	Bimal Printing Works	Bakerganj, Patna-4
6.	Bihar Press	Bakerganj, Patna-4
7.	Chitragupta Press	Makhaniya Kuan, Patna-4
8.	Chandra Prabha Press	Lohanipur, Patna-3
9.	Dayal Printing Press	Mahendru, Patna-6
10.	Gitanjali Press	Station Road, Patna-1
11.	Kalyan Art Press	Khazanchi Road, Patna-4
12.	Litho Printing Press	Subzi Bagh, Patna-4
13.	Manoj Printing Works	Subzi Bagh, Patna-4
14.	Niranjan Printing Press	Alamganj, Patna-7
15.	National Printing House	Kadam Kuan, Patna-3
16.	Prem Printing Press	Arya Kumar Road, Patna-1
17.	Raj Press	Salimpur Ahra, Patna-3
18.	Rakshpal Mudran Kendra	R.K. Avenue, Patna-3
19.	Shanti Press	Lohanipur, Patna-3
20.	Stylo Press	Rajendra Path, Patna-1
21.	Sampada Press	Balkishunganj, Patna-7
22.	Shekhar Colour Printing Press	Naya Tola, Patna-4
	2. Presses which have closed down their business	
1.	Berojgar Press	Lohanipur, Patna-3
2.	Central Press	Kadam Kuan, Patna-3
3.	Hindustan Press	Khazanchi Road, Patna-4
4.	Janvani Press	Baidyanath Bhawan, Patna-1
5.	Kanak Press	Kadam Kuan, Patna-3
6.	Lokmukh Press	Virchand Patel Path, Patna-1

1	2	3
7.	Prabhudh Press	Mahendru, Patna-6
8.	Prakash Press	Lohanipur, Patna-3
9.	Sharda Press	Guzari Bazar, Patna-8
10.	Siaram Press	Gulzarbagh, Patna-7
	3. Sick Presses	
1.	Dharamvir Press	Mahendru, Patna-6
2.	Marshal Printers	Kankerbagh, Patna-16
3.	M.F.J. Press	Kankerbagh, Patna-16
4.	New Standard Printing	Naya Tola, Patna-4
5.	Rainbow Printing Press	Nunmuhiya, Patna-7
6.	Vani Press	Bhikhanapahari, Patna-4
7.	Vir Mudranalaya	Mithapur, Patna-1
8.	Vidya Bhushan Press	Mahendru, Patna-6
	4. Presses which have refused to attend the questionnaire	
1.	Azimabad Press	Subzi Bagh, Patna-4
2.	Jagariti Press	Lohanipur, Patna-3
3.	Muneshwar Printing Press	Exhibition Road, Patna-1
4.	Rachna Press	Musallahpur, Patna-6
	5. Presses which have made more than one declaration	
1.	Bir Mudranalaya	Mithapur, Patna-1
2.	Hunkar Printing Works	Budh Marg, Patna-1
3.	Sanjay Printing Press	B.M. Das Road, Patna-4
4.	Shri Radha Press	Daldali Road, Patna-3
5.	Uma Shankar Press	Kadam Kuan, Patna-3

APPENDIX IX

S. No.	Name of Press	Location
1	2	3
	(a) Presses which could not be traced	
1.	Ajit Printing Press	Langertoli, Patna-4
2.	Arun Printing Press	Sidhsameru Nagar, Patna-7
3.	Amarbind Press	Dundi Bazar, Patna-8
4.	Ajay Art Printers	Mahendru, Patna-6
5.	Basant Printing Press	Exhibition Road, Patna-1
6.	Bharat Mail Press	Subzi Bagh, Patna-4
7.	Bihar Provincial Press	Kadam Kuan, Patna-3
8.	City Press	City Court, Patna-8
9.	International Printing Works	Lohanipur, Patna-4
10.	Jaishree Art Cottage	Bhikhanapahari, Patna-4
11.	Janbharti Press	Kadam Kuan, Patna-3
12.	Modi Press	Machhuatoli, Patna-4
13.	New Printing Press	Langertoli, Patna-4
14.	Nayan Bharat Press	Hajiganj, Patna-8
15.	Prachi Press	Kadam Kuan, Patna-3
16.	Rup Rekha Press	Station Road, Patna-1
17.	Ramkrishna Press	Sultanganj, Patna-1
18.	R.N.C. Press	R.K. Avenue, Patna-1
19.	Rana Printing Works	Patna
20.	Shobha Press	Station Road, Patna-1
21.	Swastika Printing Works	Hajiganj, Patna-8
22.	Sima Press	Station Road, Patna-1
23.	Samaj Seva Mudranalaya	Nawal Ashram, Patna-3
24.	Surendra Mudranalaya	Subzi Bagh, Patna-4
25.	The Art Press	Subzi Bagh, Patna-4
26.	Vidyasagar Mudranalaya	Lodipur, Patna-1
27.	Yugaram Press	Bakerganj, Patna-4
	(b) Dissolved Presses	
1.	Arun Printing Press	Kunkun Singh Lane, Patna-4
2.	Abhay Press	New Market, Patna-1
3.	Arbind Press	Balkrishunganj, Patna-7

1	2	3
4.	Agrawal Press	Mithapur, Patna-1
5.	Bharat Bhushan Press	Station Road, Patna-1
6.	Hind Press	Sultanganj, Patna-6
7.	Hind Press	Subzi Bagh, Patna-4
8.	Karmshil Press	B.M. Das Road, Patna-4
9.	Lipika Press	Govind Mitra Road, Patna-4
10.	Manchand Press	Digha, Patna-1
11.	Modern Press	Kazipur, Patna-4
12.	Nawjiwan Press	Bhikhanapahari, Patna-4
13.	Pustkalay Press	Tripolia, Patna-7
14.	Pandey Printing Press	Naya Tola, Patna-4
15.	Prakash Printing Press	Pirmohani, Patna-3
16.	Pratibha Press	New Market, Patna-1
17.	Rupkala Press	Exhibition Road, Patna-1
18.	Smart Press	Mahendru, Patna-6
19.	Standard Printing Press	Dariyapur, Patna-4
20.	Sangh Mudranalaya	Rajendra Nagar, Patna
21.	Shankar Press	Saidpur, Patna-4
22.	Trimurti Press	Kunkun Singh Lane, Patna-4
23.	Universal Press	Mahendru, Patna-6.
	(*c*) *Presses which have refused to provide any information*	
1.	Bihar Envelop Press	Dariyapur, Patna-4
2.	Bombay Envelop Press	Dariyapur, Patna-4
3.	Modern Printers	Langertoli, Patna-4
4.	Puspanjali Press	Lohanipur, Patna-3

APPENDIX X

A. Suggested Format of Profit and Loss Account of Patna University Press for the year ending 31st March, 19 . . .

To stock at commencement:
- Paper
- Work in progress
- Ink
- Types and Metals
- Other Materials

To Material purchased:
- Paper
- Inks
- Types and Metals
- Other Materials

To salaries and wages of regular staff :
- Compositors
- Mono-staff
- Proof Readers
- Machinemen
- Coolies
- Distributors
- Binders
- Computors and other workshop staff

To Payment for overtime and contract works:
- Proof reading
- printing
- Binding
- Others

To Power and Lighting for Workshop

To Licence Fee

To Balance c/d (cost of production):

By stock at the end:
- Paper
- Work in progress
- Ink
- Types and Metals
- Other Materials

By Balance C/d (cost of production)

By job printing for University Office

By job printing for University Departments

By job printing for Colleges

By job printing for Text Publication Fund

By job printing for others

By sales of waste paper and materials:
- Old types
- Waste metals
- Waste papers
- Other materials

By Income from investment and Fixed Deposits.

To Salaries:
Supervisory staff
Clerical staff
Peon and Darwan
To Telephone charges
Local
Trunk
To Stationery
To House Rent
To Magazines
To Charges for Printing for office
To Bank Charges
To Fire Insurance
To Contribution for Employees Provident Fund E.S.I.C.
To Bad Debts
To Contingencies:
Saffi
Oil
Sutli
Soap
Conveyance
Others
To Depreciations:
Printing Machine
Mono Equipment
Types and Metals
Cutting Machines
Stitching Machines
Furniture and Fixtures
Cycle and Loose tools
To Balance transferred to Balance-sheet Net Profit

(*Contd.*)

B. Balance-sheet of Patna University Press as at the 31st March, 19 . . .

Liabilities	*Assets*
Capital Account:	Machinery:
Payment advance from University	Printing Machines
	Less Depreciation
Add Profit during the year	Mono Equipment
Sundry creditors:	Less Depreciation
Short advance from University	Types and Metals
Text Publication Fund	Less Depreciation
Government Loan	Cutting Machines
Outstanding	Less Depreciation
	Stitching Machines
	Less Depreciation
	Loose tools and other equipments
	Less Depreciation
	Other Assets:
	Furnitures and Fittings
	Less Depreciation
	Cycle
	Less Depreciation
	Stock in Trade
	Papers
	Work in progress
	Types and Metals
	Other Materials

APPENDIX XI

Minimum Wages Act, 1948—Section 4(1)

SCHEDULE

Scale of Pay of Printing Press Employees

S. No.	*Categories of Employees*	*Minimum rates of wages*
		(*Rs.*)
1.	Manager	450
2.	Artist	360
3.	Litho Machineman	360
4.	Machineman for Offset Machine	360
5.	Flat Machineman	300
6.	Accountant	350
7.	Lino operator and Mono operator	340
8.	Cashier	340
9.	Compositor grade I	300
10.	Compositor grade II	270
11.	Typist	300
12.	Proof Reader	300
13.	Clerk	270
14.	Store-keeper	270
15.	Zinc corrector or corrector in Offset printing	270
16.	Folder	270
17.	Treadleman	270
18.	Daftari	250
19.	Binder	250
20.	Fly Boy, Paper man, Distributor, Helper and any other semi-skilled workmen	250
21.	Darwan, Peon, Chaukidar, Sweeper and any other un-skilled men	240

The minimum rates of wages are used in the All India Average Consumers Price Index number (1960-80) 425 which is the average of the indices of the first half of the year 1981 (January-June). There will be increase or decrease with rise or fall, as the case may be, in the aforesaid index at the rate of 4

Paise per point per day and that will be treated as dearness allowance component. But even if the index falls below 425, the wages shall not fall below the minimum rates of wages as shown in the schedule above. The payment of D.A. on the basis of average index of July-December, 1981 will be due from April 1982 and thereafter, increase or decrease in the amount of D.A. based on the average of the A.I.C.P.I.W.D. for the successive periods of six months each will be applicable in expiry of three months (14/RL-1026/81-L-2E-694).

By Order of the Governor of Bihar

S.P. Bose
Dy. Secretary to Government
of Bihar

Source : Bihar Gazette, letter No. 14/RL-1026/81-L-2E-694, by the order of Governor of Bihar, S.P. Bose, Deputy Secretary to Government.

APPENDIX XII

Questionnaire on Problems of Printing Industry framed and Circulated for Discussion at a Seminar of Printers, Government Officers, Academicians and Other Experts

A. Capacity Utilization

1. Is the problem seasonal or perennial?
2. What exactly is the quantum of under-utilization of capacity?
3. Why does the work go outside the State:
   (*a*) Are the Printing rates at Patna high?
   (*b*) Is the quality of work bad?
   (*c*) Are delivery promises not kept?
   (*d*) Are government rules responsible for this?
   (*e*) What can the government do about it?
   (*f*) Is some corruption involved in the process?
4. If existing capacity is being not fully utilized, why do new presses come into existence?
5. Should the establishment of new presses be banned?
6. Is the Government Press incapable of undertaking the entire government work?
7. Does the government pay for the work promptly?

B. Government and Printing Industry

1. Can the government forbid printing work being sent outside the State?
2. What will be the justification for such a step?
3. Will it be equitable to do so?
4. Will it not curtail the freedom of the people in getting their work done wherever they like?
5. Why should we plead for a separate department of the government to help the printing industry?
6. What other departments has the government started to help other industry?
7. What functions should such a department discharge?

8. Will the establishment of such a department curtail the freedom of the printing industry?
9. Will not the department be inclined to exercise control on this industry in an ever increasing measure?
10. Is not our experience of government control of business already bitter?
11. How can the government supply finance to the industry?
12. Do you need some regulatory law for the industry apart from the various laws that already exist?
13. Should the State collect periodical accounts and returns to obtain authentic information in respect of the industry?
14. Should the government assume power to distribute government and semi-government work between different presses equitably?

C. Labour in Printing Industry

1. What condition of work does the industry provide?
2. Is the location of presses suitable?
3. Are the buildings suitable?
4. Is there enough space in the presses?
5. What is the rate of turnover and absenteeism?
6. What is the state of trade unionism in this industry?
7. How far a collective bargaining is effective?
8. Do all the presses in the industry pay E.S.I. contribution?
9. How has E.S.I. Scheme affected the industry?
10. Does it have some corrupting influence on labour?
11. Is training of labour necessary?
12. Are proper training facilities available?
13. Are the training facilities satisfactory? If not, what arrangements do you suggest?
14. Does the industry face labour trouble in terms of labour shortage and labour disputes?
15. What are the causes of labour disputes?
16. What are the machineries available for the settlement of disputes in the industry?
17. Is contract system more prevalent? If so, for what type of work?

18. What are the rates generally paid for different kinds of work?
19. Is Mudrak Parishad doing something to standardise the rates?
20. How far have wages gone up during the last ten years?

D. Co-operation in Printing Industry

1. Can co-operation as a method of organisation be used for solving the problems of printing industry?
2. If yes, what are the spheres in which co-operative effort can be fruitful?
3. Can it help us in the procurement and distribution of paper metals and other materials?
4. Can a workshop for repairs be started on a co-operative basis?
5. Can equitable distribution of work be undertaken through co-operative effort and so on?

APPENDIX XIII

Questionnaire on Labour Relations

A. Printers Association

1. Is there a Printers Association?
2. Are you a member of that Association?
3. Year of Establishment
4. Name of the Association
5. Is it registered? If so:

   (*a*) Registration No.........
   (*b*) Act.........
   (*c*) Year.........

6. Location of office
7. Jurisdiction: National/State/Local
8. Source of Finance

   (*a*) Membership fee
   (*b*) Donations
   (*c*) Other sources
   (*d*) Kindly give income and expenditure a/c and Balance-sheet for the last three years: If available.

9. Membership

   (*a*) Membership fee per head
   (*b*) Total number of members

10. Any other parallel Association? If so:

   (*a*) Name
   (*b*) Date of Establishment

11. Main functions

B. Employees Association

1. Name of the Sangh
2. Year of Establishment
3. Registration..............Year..................Act...............
   No...............

4. Location of office
5. Political Affiliation with National Federation: If any
   - (*a*) A.I.T.U.C.
   - (*b*) I.N.T.U.C.
   - (*c*) H.M.S.
   - (*d*) U.T.U.C.
   - (*e*) B.M.S.
   - (*f*) C.I.T.U.
6. Membership
   - (*a*) Total number of members
   - (*b*) Membership fee per head
7. Source of Finance
   - (*a*) Membership fee
   - (*b*) Donations
   - (*c*) Other sources
   - (*d*) Kindly give income and expenditure a/c and Balance-sheet for the last three years: If available.
8. Is there any parallel Sangh? If so:
   - (*a*) Name..................Associated with............
   - (*b*) Year of Establishment.

Signature

Bibliography

Books

Bolar, Malathi, *Performance Appraisal, Readings, Case Studies and a Survey of Practices*, Vikas Publishing House Pvt. Ltd., New Delhi, 1978.

Chakraborti, M.L., *Bibliography in Theory and Practice*, The World Press Private Ltd., 2nd Ed., 1975.

Cunnings, A.J., *The Press and a Changing Civilization*, London, 1936.

Durrant, W.R., *Penrose, The International Review of the Graphic Arts*, Worthwood Publication Ltd., I.S.B.N., London, 1978.

Harrop, C., *Offset Printing from Stone Plates*, New York, 1927.

Indian Institute of Personnel Management, *Personnel Management in India, The Practical Approach to Human Relations in Industry*, Asia Publishing House, 2nd Ed., 1976.

Johan, C. Tarr, *Printing Today*, Geoffrey Cumberdege, Oxford University Press, 1949.

Marshal Johan, *Mohenjodaro and the Indus Civilization*, (Plates c ii cxvi), London, 1931.

Mestle, J.S. and Gordon, L. Monsen, *Photomechanics and Printing*, Oxford and I.B.H., New Delhi, 1950.

Montague, Noel, *Printing Management*, Sylvan Press, 2nd Ed., London, W.I., 1948.

Mudran Byabsai, *Kendrie Anusandhan tatha Rojgar Seva Sansthan, Rojgar tatha Prashikshan Mahanidesalay*, Shram Mantralay, Bharat Sarkar, Pussa, New Delhi, 1972.

Oswald, C. John, *A History of Printing*, D. Appleton & Company, New York, 1928.

Pandey, Chhabinath, *Mudran Kala*, Bihar Rashtra Bhasha Parishad, Patna, 1st Ed., 1958.

Parikh, R.D., *The Press and Society: A Sociological Study*, Popular Prakashan, Bombay, 1905.

Priolkar, Anant Kakba, *The Printing Presses in India, Its beginnings and early Development*, Bombay University Press, Fort Bombay, 1st Ed., 1958.

Ranganathan, S.R., *Physical Bibliography for Librarians*, Asia Publishing House, London, 2nd Ed. 1974.

Sahay, S.N., *Financial Management of Public Enterprises*, S. Chand & Company Pvt. Ltd., New Delhi, 1st Ed., 1984.

Saifuddin, T.A., *Management Aspect of Printing Industry*, Nirmala Sadanand Publishers, Bombay, 1st Ed., 1970.

Saxena, R.C., *Labour Problems and Social Welfare*, K. Nath & Company, Educational Publishers and Printers, Meerut, 13th Ed., 1974.

Sen, P.K., *The Press Publications and Copyright Laws of India*, M.C. Sarkar & Sons, Pvt. Ltd., Calcutta, 1958.

Singh, Ram Mohan, *The Bihar Shops and Establishment Act*, 1953, Eastern Law Book Agency, Law Book Sellers & Publishers, Patna, 2nd Ed., 1965.

Taylor, Archer and Arlt, O. Gustave, *Printing and Press*, University of California Press, Berkeley and Los Angels, 1941.

Thakur, C.P. and Munson, Fred C., *Industrial Relations in Printing Industry: A Study in the Context of Technology and Markets in Delhi Area*, Sri Ram Centre for Industrial Relations, New Delhi, 1969.

Updike, P.B., *Printing Types, Their History, Forms and Use: A Study in Survivals*, Oxford University, Published in U.S.A. by Harvard University Press, Cambridge, Massachusetts, 1937.

Whiteside, Thomas, *The Block Buster Complex, Conglomerates, Show Business, and Book Publishing*, Wesleyan University Press, Middle Town, Connecticut, U.S.A., 1st Ed., 1980.

Encyclopaedia

A New Survey of Universal Knowledge, *Encyclopaedia Britannica*, Volume 8, I.N.C. William Benton, Publisher, London, 1968.

The Encyclopaedia Americana, International Edition, American Corporation International Quarter, 575, Letington Avenue, New York, 1976.

Report and Official Publications

Government of India, *Gazetteer of India*, Bihar, Patna, Gazetteers Branch, Revenue Department, Patna, Government of Bihar, 1970.

Government of India, *Press in India*, 1968, 12th Annual Report 1969, Ministry of Information and Broadcasting, India.

Government of India, *India Annual*, 1984, Publication Division, Ministry of Information and Broadcasting, Government of India.

Government of Bengal, *Press List of Ancient Documents Relating to the Provincial Council*, at Calcutta preserved in the Secretariat Record Room of the Government of Bengal, Vol. 3, (PT), Calcutta, 1930.

Government of Bihar, *Draft Annual Plan, Data Relating to Public Sector Undertaking State Undertaking*, 1984-85.

Government of Bihar, *Income and Expenditure*, 1980-81, 1981-82, 1982-83. (Non planned), Finance Department, Government of Bihar.

Government of Bihar, *Statistical Hand Book*, 1980, Director of Statistics and Evaluation, Government of Bihar, Secretariat Press, Patna, 1983.

Government of Bihar, *Secretariat Instructions*, Government of Bihar, Revised Ed., 1952.

Government of Bihar, *Origin and Development of Bihar Legislature*, 1982.

Government of Bihar, *Bihar at a Glance, Bihar Through Figures*, Director of Statistics and Evaluation, Bihar, 1979.

Government of Bihar and Orissa, *Administrative Report*, Bihar and Orissa Government, 1911-12.

Patna University, *Report of the Committee Constituted by the Vice-Chancellor to examine the Working of the Patna University Press*, 1966.

Press India, 1969, 13th Annual Report of the Registrar of Newspapers for India under the Press and Registration of Books

Act, Ministry of Information and Broadcasting, Government of India, New Delhi, 1969.

Press in India, 1981, 25th Annual Report of the Registrar of Newspapers for India under the Press and Registration of Books Act, 1971, Part I, II, Ministry of Information and Broadcasting, Government of India, New Delhi, September, 1981.

The Bihar State Text Book Publishing Corporation Ltd., Patna, *Annual Report*, 7th and 12th, Yatin Press, Patna and Text Book Press, 1972-73 and 1976-77.

Willing's Press Guide, Seventyfifth Annual Issue, Willing's Press Service Ltd. Willing House, 356-364, Grays Inn Road, London, W.L.L. 1948.

Periodicals

(*a*) *Monthly*

Bihar Information, June 1983, (combined Issue), Vol. 31, No. 6.

Monthly Abstract of Statistics, September, 1948, Central Statistical Organisation Department, Ministry of Planning, Sardar Patel Bhawan, Sansad Marg, New Delhi, India, 1984.

Mudrak Parishad, Patna, *Printing*, A monthly magazine, March 1973, Printed at Sunil Printing Press, Patna.

Mudrak Parishad, Patna, *Mudran, Print Mag, Smarika*, Ed. 10th Annual Conference, 1972 and 1982.

Mudrak Parishad, Dhanbad, *Udyog Darpan*, The Industrial Guide, April-May, 1981.

Nirmal Brajanandan, *Print Mag*, A Mudrak Parishad Publication (Monthly, Print Mag Prakashan Samity, Rajendra Nagar, 3/35, Patna, 1984.

The Indian Worker, Special INTUC, November, 1956.

(*b*) *Weeklies*

Srivastava, Surajnath, *Udyog Bandhu*, Weekly, Sri Ram Mudranalaya, Saidpur, Patna, 1984.

(*c*) *Dailies*

The Times of India, Bombay.

The Indian Nation, Patna.

The Searchlight, Patna.

Others

Bihar Text Book Karamchari Sangh, Registration No. 2026, Patna Charter of Demands Register.

Bihar Text Book Publishing Corporation Ltd., Inauguration of the Text Book Press, Patna, India's biggest Offset Printing Complex, Text Book Press, 18th November, 1972.

Charter of Demands, Gulzarbagh, Printing Press, Karamchari Sangh.

Charter of Demands, Text Book Corporation Employees Association, Budh Marg, Patna.

Mudrak Parishad, Memorandum of Association, Patna, 1971.

Index